Atomic Physics

PHYSICS AND ITS APPLICATIONS

Series Editors

E.R. Dobbs
University of Sussex

S.B. Palmer
Warwick University

This series of short texts on advanced topics for students, scientists and engineers will appeal to readers seeking to broaden their knowledge of the physics underlying modern technology.

Each text provides a concise review of the fundamental physics and current developments in the area, with references to treatises and the primary literature to facilitate further study. Additionally texts providing a core course in physics are included to form a ready reference collection.

The rapid pace of technological change today is based on the most recent scientific advances. This series is, therefore, particularly suitable for those engaged in research and development, who frequently require a rapid summary of another topic in physics or a new application of physical principles in their work. Many of the texts will also be suitable for final year undergraduate and postgraduate courses.

1. **Electrons in Metals and Semiconductors**
   R.G. Chambers
2. **Basic Digital Electronics**
   J.A. Strong
3. **AC and DC Network Theory**
   Anthony J. Pointon and Harry M. Howarth
4. **Nuclear and Particle Physics**
   R.J. Blin-Stoyle
5. **Thermal Physics**
   Second edition
   C.B.P. Finn
6. **Vacuum Physics and Techniques**
   T.A. Delchar
7. **Basic Electromagnetism**
   E.R. Dobbs
8. **Quantum Mechanics**
   Second edition
   Paul C.W. Davies and David S. Betts
9. **Statistical Physics**
   Second edition
   Tony Guénault
10. **Photothermal Science and Techniques**
    D.P. Almond and P.M. Patel
11. **Physics in Medical Diagnosis**
    T.A. Delchar
12. **Atomic Physics**
    D.G.C. Jones

Atomic Physics

D.G.C. JONES
Honorary Lecturer in Experimental Physics
University of Sussex, UK

CRC Press is an imprint of the
Taylor & Francis Group, an **informa** business

Published by Chapman & Hall

First edition 1997

Published 2023 by CRC Press
Taylor & Francis Group
6000 Broken Sound Parkway NW, Suite 300
Boca Raton, FL 33487-2742

ISBN 13: 978-0-412-78280-0 (pbk)

A Catalogue record for this book is available from the British Library

Visit the Taylor & Francis Web site at
http://www.taylorandfrancis.com

and the CRC Press Web site at http:
//www.crcpress.com

Contents

Preface

In the past twenty years, atomic physics has moved back to the forefront of experimental and theoretical research. The ability to isolate and manipulate single atoms has made possible experiments which would once have been considered inconceivable and has enabled new tests of fundamental quantum theories.

This book is aimed at students in the first or second year of a university physics course who are meeting atomic physics for the first time, and is based on a course given at the University of Sussex between 1992 and 1995. The basic idea is to discuss the physics of the atom as a topic in its own right not, as often happens, as part of a general course in Quantum Physics, where the essence of the subject is often missed. The aim is to be as rigorous as possible, but not to submerge the student in a mass of mathematical detail.

I have ruthlessly discarded all the 'Old Quantum Theory' which is often used as a way in to the subject and which, in my opinion, serves only to obscure students' understanding. Instead, starting from the evidence for the existence of the nuclear atom, I give a very brief and (I hope) comprehensible introduction to the quantum theory of atomic particles, leading to a statement of Schrödinger's equation and the nature of its solutions for an electron in a central field. I do not labour this, but assume that students have gone through, or are going through, a basic course in quantum mechanics.

In the first half of this century, atomic spectroscopy provided the vast majority of the experimental evidence required to build up a detailed picture of atomic structure. I feel that it is necessary to sketch the outlines of this subject so that students can get a little understanding of the immense skill and dedication of the physicists who built up the huge body of experimental work on which atomic theory is based.

The central point of the book is the theory of interaction of atoms with fields, a complex topic often treated very briefly in older introductory texts. Use of Einstein's A and B coefficients can give a surprising amount of insight into the processes taking place. The quantum theory required for an understanding of the make-up of those coefficients is difficult for beginners. I hope that the derivations produced here are clear and will point the way to more detailed analyses, if required. I have also included a discussion on the modern picture of the photon, a topic often mishandled in introductory texts.

Atomic structure and the periodic table are introduced and discussed in enough detail to give a good grounding for further work. Fine structure and coupling schemes are discussed briefly, but without too much attention to their complexities. Some of the theory of laser systems that has been included might seem at first sight to be a little out of place, but lasers play such an important part in modern atomic physics that a knowledge of their properties is essential.

The chapter on interactions with external fields covers the familiar Zeeman and Stark effects, but also deals with Rabi oscillations and atomic beam experiments, subjects vital in modern experimental atomic physics. Finally, there is a very short introduction to modern fields of interest.

I have quoted many theoretical results with, I hope, enough analysis to allow the student to understand their development and over what ranges they are valid. Important derivations have been kept in appendices, to go through or leave as necessary. A detailed introduction to the topic of angular momentum is included, but in-depth questions of atomic structure are left to more advanced texts, some of which are noted in an appendix.

Most nomenclature is standard. I have added the units in more equations than is common, in order to stress the down-to-earth nature of what is, after all, an experimental subject. In line with this, all frequencies are expressed in Hz, rather than radians per second. Because of this, formulae are expressed in terms of Planck's constant h rather than 'h-crossed' ($h/2\pi$). This is purely a personal preference and, since it is consistent throughout the book, should not cause any difficulties. Problems are mostly based on ones set in my course.

Since the book is based on a lecture course, I should like to express my gratitude to those students on whom it was tested. In particular I should like to thank three excellent students from the final year in which I taught it, Ronald Baskin, Mark Camidge and Andy Waye, whose comments and (usually) constructive criticism spurred me into starting to write. My colleagues helped in many ways in discussions, although any errors or misinterpretations are, of course, my own. Roland Dobbs kindly gave encouragement. Finally, thanks to my wife Kate and son Bryn for letting me get on with it and take up a lot of time on the family computer.

Main symbols used in text

A_{nm}	Einstein coefficient for spontaneous emission
B_{nm}	Einstein coefficient for absorption or stimulated emission
$\boldsymbol{B}$	magnetic field vector
b	impact parameter
d	interatomic distance
D	classical dipole moment
D_{nm}	quantum dipole moment matrix element
$\boldsymbol{E}$	electric field vector
E	energy
E_n	energy eigenvalue associated with eigenfunction u_n
$\boldsymbol{F}$	force
H	Hamiltonian or total energy operator
I	radiation intensity
$\mathbf{l}$	quantum orbital angular momentum operator
$\boldsymbol{L}$	classical orbital angular momentum vector
m	mass of particle
$\boldsymbol{M}$	magnetic dipole moment
N	density (atoms/unit volume)
$\boldsymbol{p}$	general classical momentum vector
$\mathbf{p}$	general quantum momentum operator
$P(r,t)$	probability associated with $\Psi(r,t)$.
$\boldsymbol{P}$	classical electric polarization vector
$\boldsymbol{S}$	classical spin angular momentum vector
$\mathbf{s}$	quantum spin angular momentum operator
u_n	energy eigenfunction
υ	speed
v	volume
V	potential energy; volume of one kg-molecule
$\boldsymbol{x}$	general classical position vector
$\mathbf{x}$	general quantum position operator

Z	Atomic number
α, β	electron spin eigenfunctions
α_m, β_n	laser gain constants
γ	natural linewidth of transition
λ	wavelength
ρ	density
$\rho(\nu)$	energy density in field between frequencies ν and $\nu + d\nu$
ν	frequency (Hz)
$\Psi(r,t)$	general time-dependent wave function
$\psi(r,t)$	time-dependent wave function

1

Introduction

Although the Greek philosopher Democritus had postulated the existence of atoms in the first century BC and Dalton's atomic theory of 1807 laid the basis for modern chemistry, there was no direct evidence for the existence of atoms before the turn of the twentieth century. Indeed, at that time an influential school of German physicists led by Ernst Mach considered the atomic model to be merely a useful picture with no basis in reality.

1.1 THE EXISTENCE OF ATOMS

The situation was dramatically changed by an explosion of experimental investigation over the fifteen years between 1897 and 1912. In the 1870s, technical improvements in the construction of vacuum pumps had made possible the investigation of electrical phenomena in evacuated tubes and the discovery of invisible rays which travelled between an electrically negative electrode (cathode) and an electrically positive electrode (anode) in such a tube.

These rays came to be known as cathode rays. At first there was considerable controversy over their nature, but a series of experiments carried out by J.J. Thomson in 1897 demonstrated conclusively that the cathode rays consisted of a stream of negatively charged particles, presumably emitted by atoms in the cathode(Fig. 1.1).

Thomson's measurements of the deflection of the rays by electric and magnetic fields enabled the speed of the particles to be measured and also the ratio of the charge of a particle to its mass. By the turn of the century, the charge–mass ratio of these particles, which came to be called electrons, could be measured to quite high precision.

However, to give absolute values of the charge and mass, experiments of a different type were required. The most successful were investigations where macroscopic particles such as oil droplets were charged in some way and their motion in electric fields observed. A relatively straightforward measurement of

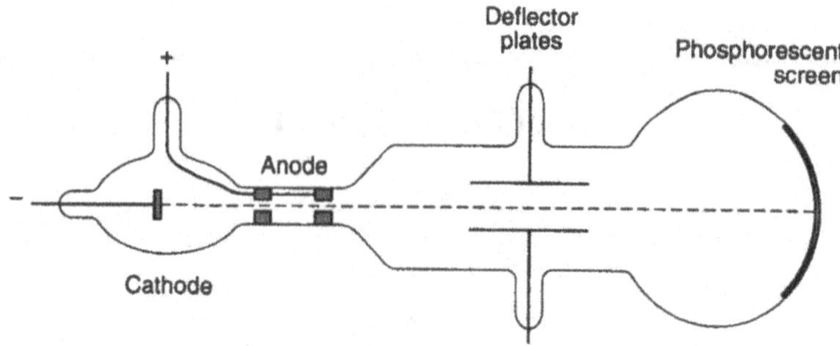

Fig. 1.1 Schematic diagram of J.J. Thomson's cathode ray tube. Electrons emitted by the cathode are accelerated through the anode. The beam of electrons hits the phosphorescent screen, producing a luminous spot.

the mass of the oil droplets enabled the charge of the electron to be measured. The famous experiments carried out by Millikan between 1909 and 1916 gave a value for this charge as $1.592 \pm .002 \times 10^{-19}$ coulomb, less than 1 percent lower than that accepted today. This, combined with Thomson's results, gave a value for the electron's mass of approximately 9×10^{-31} kg.

The measurement of electric charge made possible a direct measurement of atomic masses. Back in 1830, Faraday had carried out experiments on electrolysis. He had used his results to suggest that if matter were atomic, then electricity should also be atomic, but the converse is also true. The flow of electric current between two metallic plates in an electrolyte results in a measurable increase in the mass of one electrode. The mass of metal deposited per unit charge flowing can be measured. Assuming that the motion of atoms between electrodes is due to the fact that each atom in the electrolyte carries a specified number of excess electrons, the mass of a single atom can be calculated.

The investigation of cathode ray tubes produced another important line of experimentation. In 1895 Röntgen had discovered that cathode rays impinging on glass or metal produced a new type of ray – the X-ray. These rays were shown to have wave-like properties and in 1899 their wavelength was estimated by the Dutch physicists Haga and Wind to be of the order of 10^{-10} m, using diffraction at a v-shaped slit. In 1906 Marx demonstrated that the speed of the waves was equal to that of light to within experimental error, and it became generally accepted that X-rays were electromagnetic radiation like light, but with much shorter wavelengths.

In 1912 Laue in Germany and Bragg in England demonstrated the diffraction of X-rays by the regular pattern of atoms in a crystal lattice. These diffraction patterns gave the first direct evidence of the existence of atoms and of their sizes

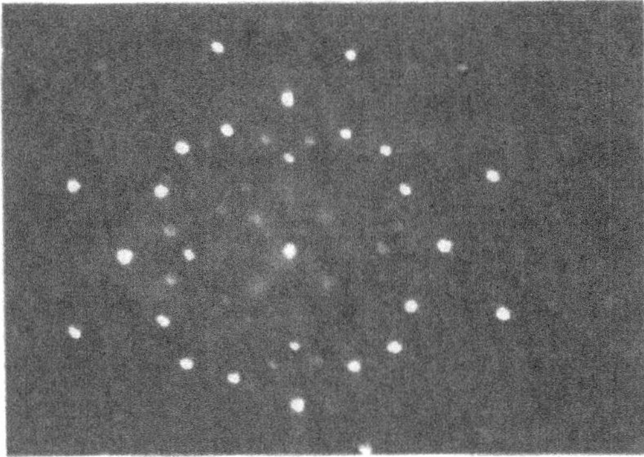

Fig. 1.2 Laue diffraction pattern caused by the diffraction of X-rays by the regular lattice of atoms in rock salt.

An example is shown in Fig. 1.2.

In 1897. Rutherford had found that pieces of the naturally occurring element uranium emitted two types of ray which were termed α rays and β rays. Both could be deflected by electric and magnetic fields and were therefore presumed to consist of charged particles. The β particles were found to have the same charge and mass as cathode ray electrons, so were assumed to be electrons. The α rays, on the other hand, were considerably more massive. Measurements of their charge and mass suggested that they consisted of helium atoms from which two electrons had been removed. This was confirmed by Rutherford and Royds in 1909, who fired α rays into a sealed and evacuated vessel and showed that helium accumulated in it. The evidence was conclusive that an α particle consisted of a helium atom from which two electrons had been removed.

This experiment also confirmed suggestions about the physical meaning of the atomic number Z. This number had been introduced to define the order of elements in the periodic table. Hydrogen had $Z = 1$, helium $Z = 2$ and so on. The identification of α particles with helium atoms suggested that Z defined the number of electrons in a particular atom.

By 1912, therefore, direct evidence existed on the mass of individual atoms and the size of these atoms. Even more interestingly, the electron appeared to be a constituent part of the atom, suggesting some internal structure.

1.2 THE SIZE OF ATOMS

Turning from the historical development of the subject, it is worthwhile to sum up the measurement of atomic masses and dimensions.

As mentioned above, direct measurement of atomic masses can be made using electrolysis. A typical electrolysis cell might consist of two copper electrodes

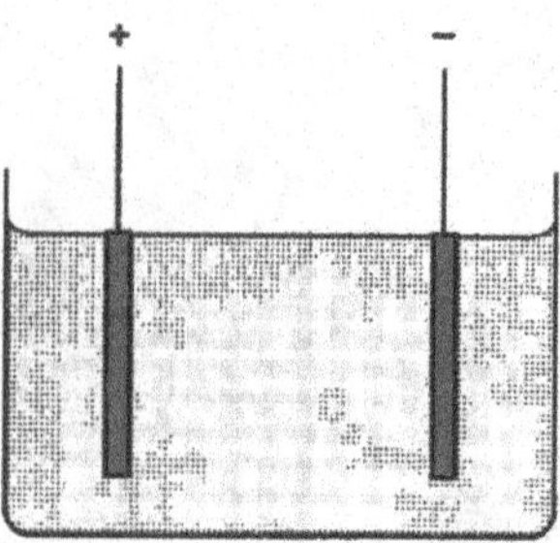

Fig. 1.3 Electrolytic cell. The anode and cathode are immersed in an electrolyte such as copper sulphate solution. Positively charged copper ions are attracted to the cathode and are deposited there.

immersed in a bath of copper sulphate (Fig. 1.3). A potential difference between the electrodes causes a current to flow and the deposition of copper on the cathode.

Several assumptions have to be made. First, it is assumed that in solution the copper sulphate crystals split up, giving free atoms of copper and that these free atoms have an excess positive charge. Second, using chemical knowledge that copper is divalent, it is assumed that the copper atom has lost two electrons. This is a reasonable extrapolation from chemical valence theory, if it is assumed that chemical bonds result from the exchange of electrons, and that the lightest atom, hydrogen, has only a single electron to exchange. A copper atom in this state is referred to as being doubly ionized, Cu^{++}. A final assumption is that all copper ions attracted to the cathode stick to it and gain further electrons to become electrically neutral again. The experiment then consists of driving a known quantity of electricity through the cell and measuring the increase in mass of the cathode.

Experiments can be carried out with different elements and results confirm the atomic theory and the theory of valence. Most interesting for our discussion is the calculation of the mass of an atom of hydrogen, the lightest element. This turns out to be 1.67×10^{-27} kg, approximately 1800 times that of an electron.

Knowing atomic masses, and the density of materials, it is straightforward to obtain values for atomic dimensions. The only problem is that unless the atoms in a sample of material are arranged in a regular pattern, the answer is not very meaningful. For crystalline substances, X-ray diffraction enables the arrangement of atoms to be discovered. The dimensions of the crystal structure can then be calculated.

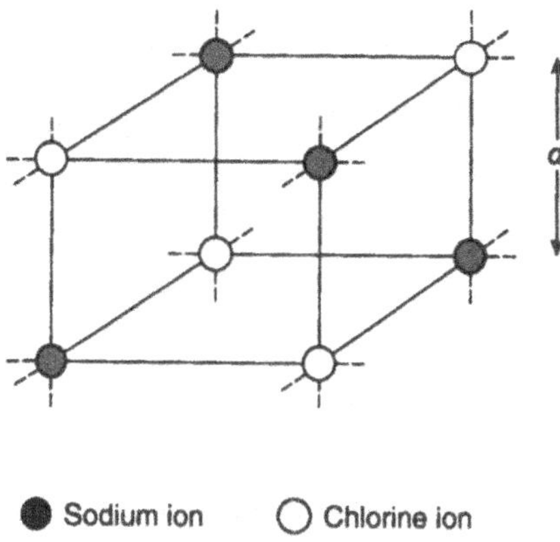

Fig. 1.4 A single cell of the simple cubic lattice of sodium chloride. The lattice is held together by the attraction between the positively charged sodium ion and the negatively charged chlorine ion.

For example, crystals of rock salt (sodium chloride, NaCl) are found to have a cubic structure, with sodium and chlorine ions on alternate corners (Fig. 1.4). If M is the kilogram molecular weight of NaCl and ρ the density of the crystal, the volume of one kg-molecule is

$$V' \quad M \quad \rho .$$

There are 2N atoms in one kg-molecule, where N is Avogadro's number. Therefore the distance between the centres of atoms, d is given by:

$$d - \sqrt[3]{(M/2\rho N)}.$$

For sodium chloride, this works out as 2.8×10^{-10} m and similar results are obtained for other crystals

Of course, such calculations only tell us the distance between the centres of the atoms and hence the maximum possible size for an atom. To go further, it is necessary to investigate the structure of the atom itself.

2

The structure of the atom

Any attempt to construct a model of the atom must start from two experimental facts. First, the atom is electrically neutral. Second, it appears to contain electrons which are negatively charged and relatively light in mass compared with the atom itself (the word 'appears' must be included, because the fact that an electron is emitted by an atom does not necessarily mean that it exists in its free-space form inside the atom).

2.1 FIRST MODELS – THE THOMSON ATOM

The first detailed model for atomic structure was devised by J.J. Thomson in 1904. Knowing the charge and mass of the electron, he used an argument from classical electromagnetic theory to estimate the radius of an electron to be of the order of 10^{-15} m, 10^{-5} times that of the atom. Thomson suggested that the atom consisted of a number of electrons immersed in a sphere of positive charge. For obvious reasons, this model is generally known as the 'plum pudding' model.

This model satisfied the major experimental criteria mentioned above. It could also explain, qualitatively at any rate, the fact that atoms emitted radiation. Lorentz had calculated that accelerated electric charges would emit electromagnetic radiation, and possible orders of magnitude of electron motion inside the atom predicted radiation frequencies of the right order. However, the Thomson model was found to be completely unable to account for new experimental results produced by Rutherford and his co-workers some seven years later.

2.2 PROBING THE ATOM

Between 1909 and 1912, Geiger and Marsden, two physicists working with Rutherford, carried out the first definitive experiments to probe the structure of the atom. The most important experiments used energetic α particles and studied

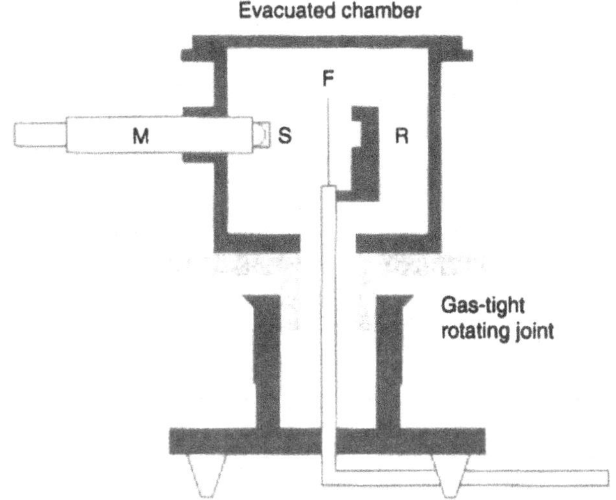

Fig. 2.1 Geiger and Marsden's apparatus for the scattering of α particles. Particles from the radium source R pass through the foil F. Scintillations at the screen S are observed via the microscope M. The screen and microscope can be rotated with respect to the foil.

their deflection as they passed through a thin metallic foil. The foil was mounted on a glass plate in a vacuum in front of an α particle source. The arrival of each particle at a zinc sulphide screen behind the sample was indicated by an observable flash of light. The screen could be rotated to observe the deflection of the particles (Fig. 2.1).

The gold foil used was less than one micron or approximately 2000 atoms thick. The α particles were known to be positively charged helium atoms with atomic number 2, and could therefore be assumed to be much smaller than gold atoms (atomic number 79). The α particles would therefore be expected to travel through the atoms and suffer an aggregate deflection which would depend on the distribution of matter within each gold atom.

Extremely surprising results were obtained. The very large majority of the particles travelled through the foil with negligible aggregate deflections. However, a small, but easily measurable proportion was deflected through very large angles and a few were found to be deflected through 180°. Rutherford described the result memorably as 'firing 16-inch shells at tissue paper and seeing them bounce back at you' No conceivable distribution of matter throughout the volume of the 'plum pudding' atom could account for such a result.

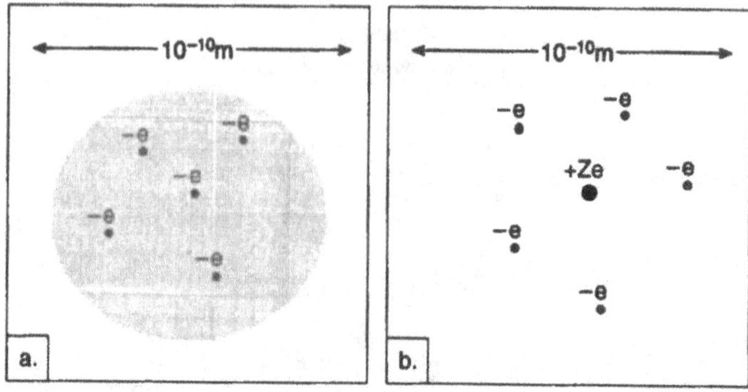

Fig. 2.2 Classical models of the atom. (a) Thomson's model. Small, negatively charged electrons are held in a dense, positively charged body. (b) Rutherford's model. The vast majority of the mass and all the positive charge are concentrated in a relatively tiny nucleus, surrounded by electrons. In both pictures the size of the electrons and of the nucleus are exaggerated. The nucleus should be at least 1000 times smaller and the electrons many times smaller again.

2.3 THE NUCLEAR MODEL OF THE ATOM

In order to explain the results, Rutherford proposed a new model in which all the positive charge and most of the mass of the atom resided in a central nucleus, surrounded by electrons orbiting in free space. The size of the nucleus would be small compared with the size of the atom (Fig. 2.2(b)). This model would give a qualitative explanation for Geiger and Marsden's results as most of the α particles would pass through the atom without encountering any matter, but a very few would collide with the massive nucleus. However, much more importantly, this model gives a precise quantitative agreement between theory and experiment.

Because of the seminal nature of this model, it is worthwhile looking at Rutherford's analysis in detail. Only classical physics is required.

The analysis of the scattering experiment falls into two parts. First, it is necessary to obtain an expression for the deflection of a single α particle as a function of its kinetic energy and its trajectory relative to the nucleus. The particle and the nucleus are assumed to be very small, and the nucleus is assumed to have a positive charge Ze where e is the electronic charge and Z the atomic number. The α particle has a charge of $+2e$ and the force between it and the nucleus is given by Coulomb's law. Figure 2.3 shows the situation, with the nucleus situated at the origin. The α particle starts far enough away from the nucleus for the interaction force to be negligible and travels parallel to the x-axis. An important parameter of the motion is the impact parameter, b, which

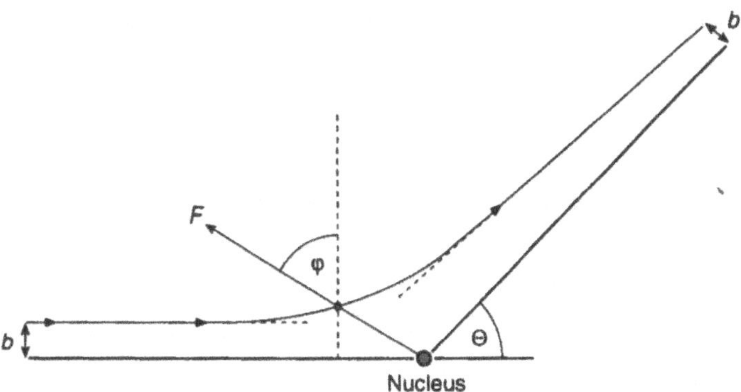

Fig 2.3 Path of an α particle (charge +2*e*) in the field of the nucleus (charge +*Ze*). The nucleus is at the origin and is very much more massive than the α particle. The force *F* is due to electrostatic repulsion.

defines the minimum distance between the nucleus and the particle if the particle were not deflected. Electostatic repulsion means that the particle is deflected through an angle Θ and it is obvious that the smaller the value of *b*, the greater is the value of Θ.

It is now possible to work out a value for Θ in terms of *b* and the kinetic energy of the particle *T*. Since the mass of the nucleus is much greater than that of the α particle, the kinetic energy and hence the speed of the particle before and after deflection remains the same. However the particle's direction of motion has changed and the law of conservation of momentum gives an expression for the absolute value of the change in momentum (Fig. 2.4)

$$\Delta p = |p_2 - p_1| = 2m\upsilon\sin(\Theta/2), \tag{2.1}$$

where m is the mass of the particle, and υ its speed.

From Newton's second law, this change of momentum must be equal to the force acting on the particle, integrated over the whole time that the particle is in the field of the nucleus. Therefore,

$$\Delta p = 2m\upsilon\sin(\Theta/2) = \int_0^\infty F dt. \tag{2.2}$$

Figure 2.3 shows the direction of *F* for a particular position of the particle, defined by the angle φ, as shown. By symmetry, it can be seen that the integral

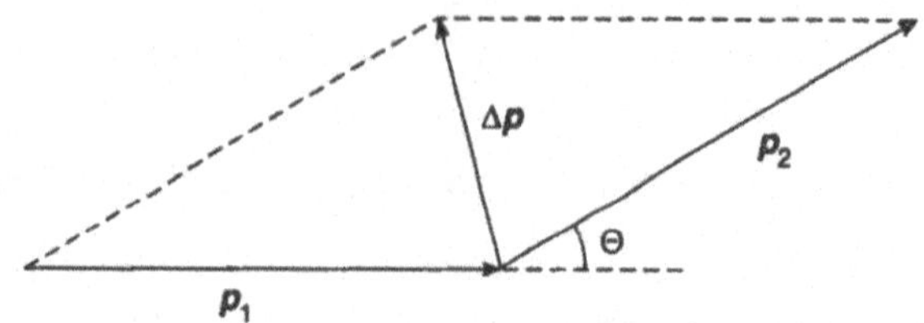

Fig 2.4 Change in momentum of an α particle during interaction with the nucleus.

in (2.2) is given by

$$I = \int_0^\infty F \, dt$$

$$= \int_0^\infty F \cos \varphi \, dt$$

(since the integral of the component parallel to the x-axis, $F \sin \varphi$, must be zero, by symmetry).

A change of variables for integration enables (2.2) to be rewritten:

$$2m\upsilon \sin(\Theta/2) = \int_{-(\pi-\Theta)/2}^{(\pi-\Theta)/2} F \cos \varphi \, (dt/d\varphi) \, d\varphi \tag{2.3}$$

(see Fig. 2.3 for the changed limits of integration).

Finally, $(dt/d\varphi)$ is equal to $1/\omega$ where ω is the angular speed of the particle about the origin. Since the force acting on the particle is radial, the angular momentum of the particle is the same for any value of φ, and ω must be given by the equation

$$mr^2\omega = m\upsilon b.$$

Therefore

$$(dt/d\varphi) = r^2/\upsilon b.$$

Coulomb's law gives

$$F = 2Ze^2/4\pi\varepsilon_0 r^2$$

so that substituting in (2.3) and integrating the right hand side gives an expression for Θ in terms of υ and b

$$\cot(\Theta/2) = (2\pi\varepsilon_0 \, m\upsilon^2 / Ze^2) \, b \tag{2.4}$$

or, in terms of the kinetic energy T of the particle

$$\cot(\Theta/2) = (4\pi\varepsilon_0 T / Ze^2) \, b \tag{2.5}$$

This gives an equation for the scattering angle in terms of the kinetic energy and impact parameter of the particle and of the charge on the nucleus, Ze.

The second part of the analysis is now required. Experimentally, although T is constant for all particles in the beam, the value of b is not. However, (2.5) can be used to calculate the fraction of particles which are deflected to some angle greater than a chosen value of Θ, say Θ_1. Figure 2.5 shows a perpendicular view of the foil with the atomic nuclei represented by points. For a particle to be scattered to an angle greater than Θ_1, it must come closer to the nucleus than the distance b_1, where b_1 is specified by equation (2.5). Assuming the particles fall uniformly on the foil, the fraction deflected to angles greater than Θ_1 is given by

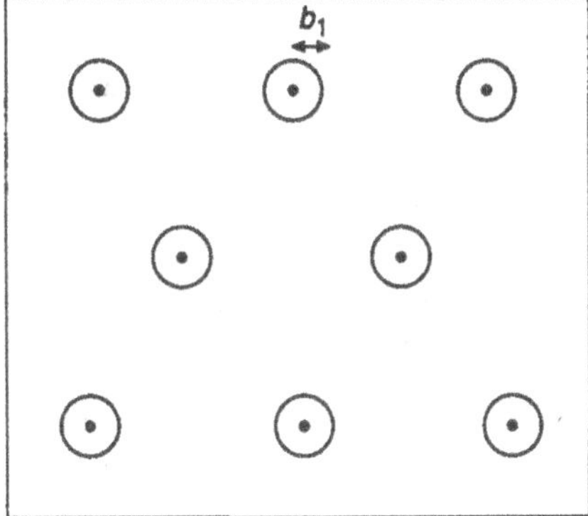

Fig. 2.5 Schematic picture of gold foil. The atoms are represented by dots. If an α particle passes closer to the nucleus than distance b_1, it is scattered to an angle greater than Θ_1, as defined in (2.5). The fraction of all particles scattered to angles greater than Θ_1 is given by the ratio of the total areas of circles of radius b_1 to the area of the foil.

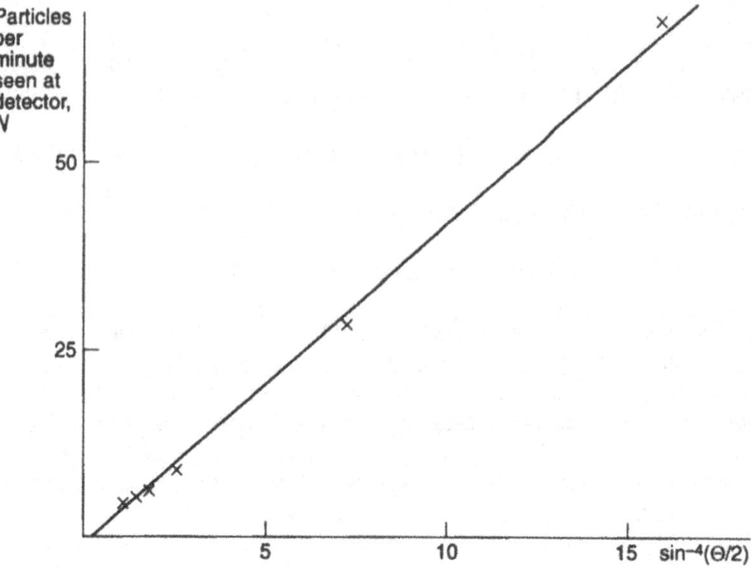

Fig. 2.6 Geiger and Marsden's results for the scattering of α particles by gold foil (plotted from the table in their original paper). The crosses signify experimental measurements and the straight line shows the theoretical variation given by (2.7).

the fraction which hit the foil within the circles of radius b_1 surrounding the nuclei. Given a foil of thickness t and area A, with a density of atoms N/m^3, this fraction f_1 is given by

$$f_1 = (\pi N t\, b_1{}^2 A)/A = \pi N t\, b_1{}^2.$$

From (2.5), this can be rewritten to give the general expression for the fraction of atoms scattered to angles greater than Θ:

$$f = \pi N t\, (Ze^2/4\pi\varepsilon_0 T)^2 \cot^2(\Theta/2). \qquad (2.6)$$

2.4 COMPARISON OF THE MODEL WITH EXPERIMENT

Equation (2.6) gives a relation which can be experimentally tested. To make measurement easier, the equation can be used to calculate the fraction of particles scattered to angles between Θ and Θ+dΘ. A straightforward piece of mathematics (see Appendix A) gives a value for $F(\Theta)$, the fraction falling on area dS at a distance r from the foil, perpendicular to r and subtending an angle dΘ at the foil

$$F(\Theta) = \pi Nt\,(Ze^2/\,8\pi\varepsilon_0 r\,T)^2\ \ [\,1/\sin^4(\Theta/2)\,]. \tag{2.7}$$

This formula, known as the Rutherford scattering formula, enables a direct comparison of Rutherford's model with the experimental results.

The comparison, using Geiger and Marsden's original results is shown in Fig. 2.6. There is excellent agreement between theory and experiment which gives very strong support for the nuclear model of the atom.

Scattering experiments gave further information about the model. Using foils of copper, gold and platinum, Rutherford's student Chadwick demonstrated experimentally that the value of Z, the measure of the positive charge on the nucleus was equal to the atomic number of the element, hence confirming the model of the atom as a massive nucleus carrying all the atom's positive charge, surrounded by Z negatively charged electrons.

Finally, an estimate of the size of the nucleus could be made. Since some α particles were seen to rebound back along their original path, they must have impinged directly on the nucleus ($b = 0$), come to rest and then rebounded. At the distance of closest approach d_0, all the particle's kinetic energy must have been converted into potential energy. The expression for the electrical potential energy of one charge in the field of another gives a value for d_o

$$d_0 = Ze^2/\,(2\pi\varepsilon_0\,T)\ \ \text{m}. \tag{2.8}$$

This gives a maximum value for the size of the gold nucleus of around 10^{-14} m, ten thousand times smaller than the size of the atom.

2.5 THE IMPOSSIBILITY OF THE CLASSICAL NUCLEAR ATOM

Rutherford's nuclear model fitted exceptionally well with the experimental data accumulated from scattering experiments. However there are great problems with it. First, since it is assumed that only electrical forces are involved in holding the structure together (gravitational attraction between the particles is negligible in comparison) no stable static structure can exist, as the electrons would be immediately attracted towards the surface of the nucleus.

A seductive alternative would be for the electrons to move in closed orbits around the nucleus – the 'sun and planet' model often used as a symbolic representation of the atom. In this case, as with the solar system, the attractive force between electron and nucleus would keep rotating electrons in a stable orbit. Also, given the size of the atom and the electronic charge, the electron's rotation frequency would be of the order of 10^{15} Hz – of the order of magnitude of the frequency of radiation emitted by atoms in discharge tubes.

However, there is a fatal flaw in this model. An orbiting electron would be always be undergoing acceleration in the direction of the nucleus and classical electromagnetic theory states unequivocally that an accelerated charged particle must emit energy in the form of electromagnetic radiation.

From electromagnetic theory, the power P radiated by a rotating charged particle is given by

$$P = 2e^2 a^2 / 12\pi\varepsilon_0 c^3 \quad \text{watts.} \tag{2.9}$$

The acceleration a of the electron is directed towards the nucleus and is given by:

$$a = Ze^2 / 4\pi\varepsilon_0 m r^2 \quad \text{m/s}^2. \tag{2.10}$$

Finally, the energy of the electron E (kinetic plus potential) is given by:

$$E = - Ze^2 / 8\pi\varepsilon_0 r \quad \text{J.} \tag{2.11}$$

As the electron emits radiation, it must lose energy. The only way E can decrease is by r decreasing, so radiation causes the electron to spiral in towards the nucleus. However, as r decreases, a increases and the law of conservation of energy enables the relationship between E and P to be written

$$P = - dE/dt.$$

Substitution from (2.9), (2.10) and (2.11) and integration gives a value for the time τ taken for an electron starting at the known atomic radius, r_a to spiral into contact with the nucleus, radius r_n, and hence for the atom to collapse in on itself

$$\tau = 4\, m^2c^3(4\pi\varepsilon_0/e^2)^2(r_a^3 - r_n^3) \quad \text{s.}$$

This time turns out to be of the order of 10^{-10}s ! In other words, Rutherford's classical atom would only exist for a very short time indeed.

There is also a subsidiary problem. The radiation emitted by electrons spiralling towards the nucleus would be continuously varying in frequency. This is not observed and, as will be seen in Chapter 4, radiation from excited atoms comprises 'spectral lines' of very well defined frequencies. However this problem is almost unimportant compared with the basic paradox. On one side, scattering experiments unequivocally confirm Rutherford's model. On the other side, this model cannot possibly be stable.

There is no way out of this problem in classical physics. No picture of a classical electron in a stable orbit around a nucleus is physically viable. In order to resolve this impasse it is necessary to reconsider the nature of the electron.

3

The quantum mechanical picture of the atom

The paradox of the nuclear atom can only be resolved in one way. Experimental evidence demonstrates conclusively the existence of a central massive nucleus surrounded by free electrons, where the sizes of both the nucleus and of the electrons are tiny compared with the size of the atom, but classical physics shows such an atom to be unstable. Therefore, the laws of classical physics cannot completely explain the structure of objects of atomic dimensions.

3.1 THE ELECTRON AS A QUANTUM PARTICLE

In fact there is direct evidence that the electron does not behave like a classical particle. Experiments which were first carried out by Davisson and Germer and by G.P. Thomson (son of J.J.), showed 'wave-like' properties of an electron beam. G.P. Thomson's experiment involved a modified cathode ray tube where electrons travelled through a thin film of graphite on their way to the fluorescent screen. What appeared on the screen was a pattern of circles surrounding a central spot (Fig. 3.1). This could not be explained in terms of the scattering of classical particles and suggested some form of diffraction process. Knowledge of the crystal structure and spacing of graphite enabled quantitative measurements to be made.

It turned out that the electrons seem to be guided by some form of wave motion, where diffraction of the wave by the regular lattice of carbon atoms in the graphite film produces the pattern on the screen. The wavelength of this wave is related to the momentum (and hence the kinetic energy) of the electrons by the simple equation

$$m\upsilon = \mathrm{h} / \lambda \qquad (3.1)$$

where m is the electron mass, υ its speed, h Planck's constant and λ the wavelength.

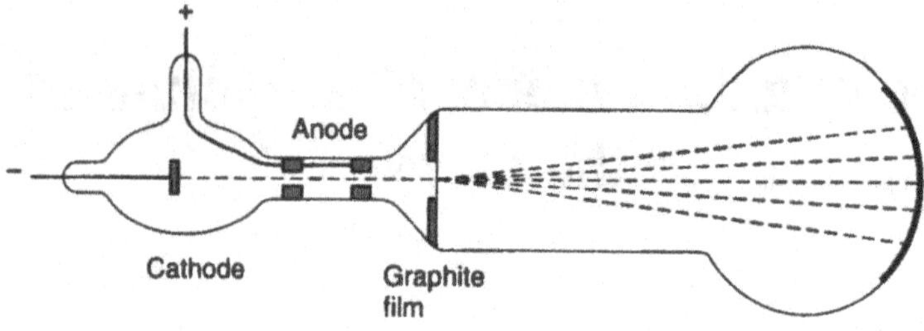

Fig. 3.1 Modification of J.J. Thomson's cathode ray tube. The electrons pass through a graphite film and a diffraction pattern of well-defined rings is seen on the screen. The ring spacing varies linearly with electron momentum (and hence as the square root of anode–cathode voltage). This leads directly to (3.1).

The important result is that the electron is not governed merely by the laws of classical mechanics and electrodynamics, but by the laws of quantum mechanics. In practical terms, this means that the motion of the electron cannot be specified exactly like that of a classical particle. The basic theory of quantum mechanics is covered in many excellent textbooks (see Appendix I). The elements necessary for the study of atomic physics are outlined in this chapter.

In classical mechanics, we deal with measured variables. For a classical particle, two general variables, position (x) and momentum (p), are needed to specify its motion and Newton's laws may be used to calculate a functional relationship for the two in terms of the force fields in which the particle moves. In quantum mechanics, on the other hand, the assumption is made that the operation of making a measurement may disturb the system being measured and that, therefore, the order in which measurements are made is important. The classical variables x and p are replaced by quantum **operators** $\mathbf{x}$ and $\mathbf{p}$. The importance of the order of measurement is expressed by the fact that the two operators do not necessarily commute, that is

$$\mathbf{x} \cdot \mathbf{p} - \mathbf{p} \cdot \mathbf{x} \neq 0.$$

In fact, a postulate of quantum mechanics is the equation

$$\mathbf{x} \cdot \mathbf{p} - \mathbf{p} \cdot \mathbf{x} = \mathrm{i}h/2\pi. \tag{3.2}$$

Other classical functions such as angular momentum, l and energy, H may be defined in terms of x and p. Similarly, quantum operators **l** and H are defined in terms of **x** and **p**. These operators will be met later.

On what do the operators operate? The state of a system in space and time is defined by a function $\Psi(x,t)$ which, for reasons which will be seen shortly, is often known as a wave function. This contains the information about the motion of the particle but this information is drastically different from that available for a classical particle. For a quantum mechanical particle, it is only possible to work out the probability of finding the particle at a particular position defined by the vector $\boldsymbol{r}$ at time t, $P(\boldsymbol{r},t)$. This is related to $\Psi(\boldsymbol{r},t)$ by

$$P(\boldsymbol{r},t) = |\Psi(\boldsymbol{r},t)|^2. \tag{3.3}$$

The function $\Psi(\boldsymbol{r},t)$ may be complex, but $P(\boldsymbol{r},t)$, being a probability, is always real.

3.2 SCHRÖDINGER'S EQUATION

In classical mechanics, the dynamic properties of a particle may be calculated using Newton's laws, particularly Newton's second law. An equivalent calculation can be made using quantum operators. A choice of expressions for the operators **x** and **p** which satisfies equation (3.2), leads to an equation known as Schrödinger's Equation (the choice normally made is the replacement of the operators **x** and **p** by the functions x and $-i(h/2\pi)\nabla$. The reasons for this are discussed in quantum mechanics texts). The solution of Schrödinger's equation gives all the information possible for systems of the size of atoms.

The general form of Schrödinger's equation for a particle of mass m moving in a force field is given below, where V is the (classical) potential energy which the particle possesses by virtue of being in the field. This may be due to any force or forces acting on the particle, but for electrons it is invariably due to electric forces. V may be a function of position and time, or may be zero.

In three dimensions Schrödinger's equation is written as

$$-i(h/2\pi)\ \partial\Psi/\partial t = (h^2/8\pi^2 m)\ \nabla^2\Psi\ -\ V\Psi \tag{3.4}$$

where (in cartesian co-ordinates)

$$\nabla^2\Psi = \partial^2\Psi/\partial x^2 + \partial^2\Psi/\partial y^2 + \partial^2\Psi/\partial z^2.$$

For general discussion, it is easier to consider a particle moving in the x-dimension only, and rewrite the equation

$$-ih\ \partial\Psi/\partial t = d^2\Psi/dx^2. \tag{3.5}$$

This equation has the same mathematical form as a wave equation, where Ψ is equivalent to the wave function. This suggests an answer to the puzzle of the 'wave-like' properties of the electron.

Because of this, the Ψ functions are sometimes called 'probability waves'. They do not have a physical existence like electromagnetic or sound waves, but have the same mathematical form. This is why $\Psi(x,t)$ is known as the 'wave function', with equation (3.3) giving its physical interpretation. Although Ψ may be positive, negative or complex, the value of P, representing a probability, will always be positive. Furthermore, the integral of P over all space and time must be unity, since the particle must exist somewhere at some time, so the value of Ψ may be 'normalized' by use of the equation

$$\int_{-\infty}^{\infty} |\Psi(x,t)|^2 \, dx dt = 1. \quad (3.6)$$

There is one important special case which will come up again and again. If a particle is moving in a field which itself does not vary in time (V in (3.4) or (3.5) is independent of time), Ψ may be separated into the product of two terms

$$\Psi(x,t) = u(x)\, f(t). \quad (3.7)$$

Substituting equation (3.7) in (3.5) enables the equation to be separated out as shown below

$$-i(h/2\pi f)\partial f/\partial t = [(h^2/8\pi^2 m)d^2u \; dx^2 - Vu]\; u. \quad (3.8)$$

The left hand side of this equation depends only on t, while the right hand side depends only on x. For consistency, both sides can therefore only be equal to some constant E. Solving for the time-dependent part, it turns out that

$$f(t) = \exp\,(-i2\pi Et/h).$$

and E has units of energy. Equations (3.3) and (3.4) may now be written

$$P(x) = |u(x)|^2$$

and

$$\int_{-\infty}^{\infty} |u(x)|^2 \, dx = 1. \quad (3.9)$$

The spatially-dependent part of equation (3.8) is known as the time-independent Schrödinger equation (TISE for short) and has the form

$$-(h^2/8\pi^2 m)\, d^2u(x)/dx^2 + Vu(x) = Eu(x). \quad (3.10)$$

The solution of this equation gives the value of the function $u(x)$. From that solution, the probability of finding the particle at any particular position x can be obtained.

In other words, the probabilities of observing a particle at two positions x_1 and x_2 are given by $|u(x_1)|^2$ and $|u(x_2)|^2$. Note that Schrödinger's equation cannot tell us **anything** about the way in which the particle may have travelled between x_1 and x_2. It cannot define the trajectory of a particle in the way that Newton's laws can.

It is worth noting how (3.10) ties in with the discussion of operators given earlier. In terms of the operators **x** and **p**, the left hand side of the equation is merely the total energy of the particle. So the equation may be rewritten as

$$\mathrm{H}\, u(x) = E\, u(x) \tag{3.11}$$

where H is the **total energy operator** or **Hamiltonian operator**.

Usually, there will be more than one solution for (3.10) (or (3.11)). The set of solutions is known mathematically as the set of **eigenfunctions** of the equation (mathematically, (3.11) is often known as an eigenequation). For each eigenfunction u_n, there will be an equivalent **eigenvalue** E_n which will define the energy of the particle which exists in that particular eigenfunction. The set of eigenfunctions will obviously depend on the form of V and will also depend on the boundary conditions imposed on the equation.

The eigenfunctions have the important property of orthonormality. That is

$$\int_{-\infty}^{\infty} u_n^*(x)\, u_m(x)\mathrm{d}x \begin{cases} = 1 \text{ for } n = m \\ = 0 \text{ for } n \neq m \end{cases}$$

where $u_n^*(x)$ is the complex conjugate of $u_n(x)$.

The word 'eigenfunction' comes from the German. The nearest English equivalent might be 'stationary function'. A simple classical analogy is the motion of waves on a wire held at both ends. For particular values of wavelength, standing waves can exist. This set of standing waves forms a set of eigenfunctions for the wave motion on the wire. Importantly, any possible motion of the wire may be made up of a linear combination of the standing waves. Similarly, it will be seen that all states of a quantum system can be made up of a linear combination of the eigenfunctions of the system.

Most calculations of the physical properties of a quantum system will consist of solving special cases of (3.10) to find sets of eigenfunctions and their equivalent eigenvalues. The eigenvalues, having units of energy, will define the possible energy states in which the particle may exist. The states may form a continuum or a discrete set. In this second case the separation between different energy eigenvalues will usually be very small compared with macroscopic energies so that at a large scale the energy of the particle may appear to be continuously variable. This may be thought of as the 'classical limit', where the quantum mechanical and classical results agree.

Thus for large-scale phenomena, classical laws still obtain, and it is often useful to consider classical 'pictures' of the properties of an electron or electrons using classical variables before converting to the equivalent quantum mechanical operators. However, a too rigid attempt to tie a classical picture to an experimental phenomenon can often produce great difficulties.

The principles outlined above will be applied to atomic systems in the following chapters. Only an outline of the necessary quantum mechanical calculations will be given. Detailed derivations of formulae and equations can be found in the appendices, where required, and in the quantum mechanics texts given in Appendix I.

3.3 AN ELECTRON IN THE ELECTRIC FIELD OF THE NUCLEUS

To apply quantum mechanics to the nuclear atom, it is necessary to write down Schrödinger's equation for this special case. For simplicity we deal first with the motion of a single electron in the field of a nucleus, which is assumed have a charge $+e$ and to be very massive and hence stationary at the origin of the co-ordinates. The potential energy of the electron is given by the classical formula

$$V = -e^2/4\pi\varepsilon_0|\mathbf{r}|. \tag{3.13}$$

V is independent of time, but the motion of the electron is in three dimensions, so the three-dimensional version of the TISE is used. V is also spherically symmetric, so it is easiest to write the equation in terms of spherical polar co-ordinates (r,θ,ϕ)

$$-(h^2/8\pi^2m)\nabla^2u(r,\theta,\phi) - e^2u(r,\theta,\phi)/4\pi\varepsilon_0|\mathbf{r}| - Eu(r,\theta,\phi). \tag{3.14}$$

Before solving this equation, it is worth recalling its physical meaning. Returning to (3.11), it can be seen that the terms on the left hand side are essentially the kinetic and potential energies of the electron and so the equation as a whole is merely a statement of the law of conservation of energy, where E defines the total energy of the electron. So the eigenfunctions will give a set of allowed probability distributions for the electron, and their associated eigenvalues will give the equivalent allowed energies.

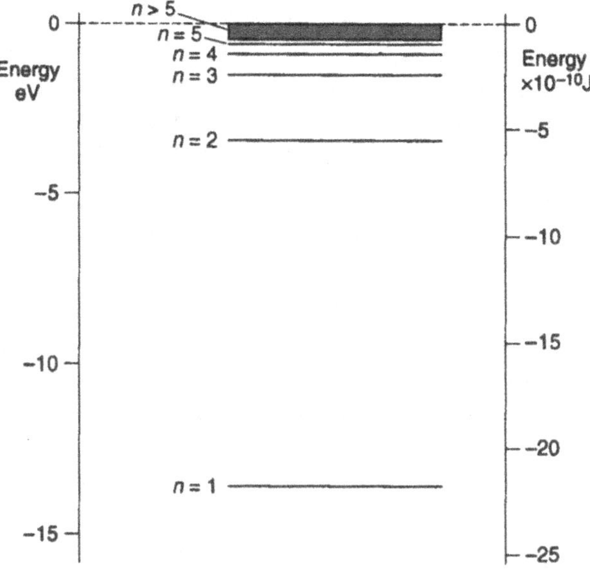

Fig. 3.2 Energy eigenvalues for an electron in the field of a massive nucleus of charge $+e$. Energies are shown in joules and in electron-volts. Levels for values of $n > 5$ cannot be resolved on the diagram.

To solve the equation, the fact is used that V is spherically symmetric. The function $u(r,\theta,\phi)$ can be separated into two terms

$$u(r,\theta,\phi) = R(r)\ Y(\theta,\phi). \tag{3.15}$$

Substitution in equation (3.14) produces two equations, one with variable R and the other with Y, and these can be solved separately.

Considering first the equation for R, the boundary condition that must be satisfied is that R must go to zero as r goes to infinity. This makes obvious physical sense in the light of (3.9).

The first and most useful result obtained from the solution is a set of values for the energy eigenvalues E. It turns out that E can only have certain values specified by the equation

$$E_n = -[me^4/8\varepsilon_o{}^2h^2] \mid 1/n^2 \mid \text{ J.} \tag{3.16}$$

The quantities within the first squared brackets are all known constants; n is an integer which may have values 1,2,3....

Physically, this gives a set of possible values for the energy states of the electron (Fig. 3.2). The state where n =1 will be the lowest possible energy state which is normally known as the ground state. Obviously, the integer n is of considerable importance and is referred to as a 'quantum number'.

3.4 ELECTRON EIGENFUNCTIONS

Evaluating the set of eigenfunctions that go with the eigenvalues is more difficult. The equation for R provides a set of solutions for the radial part of the eigenfunction which depend on the integer n. However, they do not merely depend on n, but also on a second quantum number l, which may take a series of different values, depending on n. For a given n, l may take all integral values between zero and $(n - 1)$. This means that, for example, when $n = 1$ there is only one value of l (= 0) and hence one expression for the radial component of the eigenfunction. However, when $n = 2$, there are two possible values of l (0 and 1), giving two possible expressions for the radial component, and so on for higher values of n.

The solution for the angular part $Y(\theta,\phi)$ of the eigenfunction is similarly complicated. In this case the boundary conditions are given by the cyclic nature of the function ($Y(\theta + 2\pi) = Y(\theta)$ etc.). As with the radial part, a set of solutions is obtained, defined by the two quantum numbers l and m. The quantum number l is the same one defined in the radial part, but for any given value of l, m may take all possible values l, $(l - 1)$, $(l - 2)$....0.... $-(l + 1)$, $-l$.

Putting the two parts together, each eigenvalue of u is defined by the three quantum numbers n, l and m:

$$u_{n,l,m}(r,\theta,\phi) = R_{n,l}\,(r)\; Y_{l,m}(\theta,\phi). \tag{3.17}$$

Table 3.1 gives expressions for $u_{n,l,m}$ for values of n up to 3 and a pattern can be seen. Y is constant for all states where $l = 0$, and so the eigenfunctions are spherically symmetric. Values of Y for all states where $l = 1$ have a dependence on θ which is of the form of sin θ or cos θ. Values of Y for all states where $l = 2$ have a dependence on θ which is of the form $\sin^2\theta$, $\cos^2\theta$ or $(\sin\theta\cos\theta)$. Similarly the dependence on ϕ is of the form exp $(im\phi)$.

Table 3.1 Energy eigenfunctions for an electron in the field of a nucleus of charge $+e$ for values of n up to 3. r_0 is the Bohr radius, $h^2\varepsilon_0/\pi m e^2$.

Quantum numbers (n, l, m)	$u_{n,l,m}(r,\theta,\phi)$
$n = 1,\ l = 0,\ m = 0$	$\frac{1}{\sqrt{\pi r_0^3}} \exp(-r/r_0)$
$n - 2,\ l - 0,\ m - 0$	$\frac{1}{\sqrt{32\pi r_0^3}}(2 - r/r_0)\exp(-r/2r_0)$
$n = 2,\ l = 1,\ m = 0$	$\frac{1}{\sqrt{32\pi r_0^3}}(r/r_0) \exp(-r/2r_0)\cos\theta$
$n = 2,\ l = 1,\ m = \pm 1$	$\frac{1}{\sqrt{64\pi r_0^3}}(r/r_0)\exp(-r/2r_0)\exp(\pm i\phi)\sin\theta$
$n = 3,\ l = 0,\ m = 0$	$\frac{1}{81\sqrt{3\pi r_0^3}}(27 - 18r/r_0 + 2r^2/r_0^2)\exp(-r/3r_0)$
$n = 3,\ l = 1,\ m = 0$	$\frac{\sqrt{2}}{81\sqrt{\pi r_0^3}}(6 - r/r_0)\, r/r_0 \exp(-r/3r_0)\cos\theta$
$n = 3,\ l = 1,\ m = \pm 1$	$\frac{1}{81\sqrt{\pi r_0^3}}(6 - r/r_0)\, r/r_0 \exp(-r/3r_0)\exp(\pm i\phi)\sin\theta$
$n = 3,\ l = 2,\ m = 0$	$\frac{1}{81\sqrt{6\pi r_0^3}}(r^2/r_0^2)\exp(-r/3r_0)(3\cos^2\theta - 1)$
$n = 3,\ l = 2,\ m = \pm 1$	$\frac{1}{81\sqrt{\pi r_0^3}}(r^2/r_0^2)\exp(-r/3r_0)\exp(\pm i\phi)\cos\theta\sin\theta$
$n = 3,\ l = 2,\ m - \pm 2$	$\frac{1}{162\sqrt{\pi r_0^3}}(r^2/r_0^2)\exp(-r/3r_0)\exp(\pm 2i\phi)\sin^2\theta$

3.5 THE PHYSICAL MEANING OF THE QUANTUM NUMBERS *n*, *l*, *m*.

The physical significance of the eigenfunctions of the wave function u is embodied in (3.3) and (3.9). The square of the modulus of the eigenfunction gives the relative probability of observing the electron at any point in space when it is in that particular eigenstate. Physically, this can be thought of in terms of a 'distribution' of negative charge in the space around the nucleus, so plotting the values of $|u|^2$ for the different eigenstates gives a picture of the 'charge distribution' for the eigenstate in question. This is difficult to visualize, but an attempt has been made in Fig. 3.3. By way of reference to the older picture of electrons moving in 'orbits' around the nucleus, these eigenfunctions, describing the probability distribution of an electron in the space around the nucleus, are often referred to as 'orbitals'.

It can be seen that only states where $l = 0$ have a spherically symmetric distribution and might therefore be thought of in the terms of "orbiting" electrons. All other states have a more or less complicated angular symmetry. There are two important points to note. States of given n all have the same energy, given by (3.16). Furthermore, if, for a given n, the spatial distributions equivalent to all possible values of l and m are superimposed, a spherically symmetric charge distribution results.

As far as the individual quantum numbers are concerned, n defines the energy of the state and is hence known as the principal quantum number. The angular equation for l, on the other hand is found to be identical with the general equation defining the angular momentum of a particle, so l is taken to define the angular momentum of the electron and is called the angular momentum quantum number. Finally m, which plays no part in defining either the energy or the spatial symmetry of the eigenstate, will be found later to define the possible orientation in space of the eigenstate. Because this orientation is only of importance when an atom is in a situation where an external co-ordinate system is defined, for example by an external magnetic field, m is often called the magnetic quantum number.

There is one point on which the reader may feel cheated. The reason for developing the quantum theory was the impossibility of a classical electron resting in a stable orbit without emitting radiation. But the quantum theory defines stable energy eigenstates for electrons without reference to radiation. One possible answer is that the solution refers only to possible spatial locations of electrons, and not in any way to the electron motion. However, since the question of angular momentum has been brought in, there is an assumption that the electron will be moving relative to the nucleus.

The only answer at this stage is the statement that, by definition, energy eigenstates are states of constant energy and that an electron in such a state cannot lose energy in any way. However, when the interaction of atomic systems with electromagnetic radiation is considered in Chapter 6, it will be

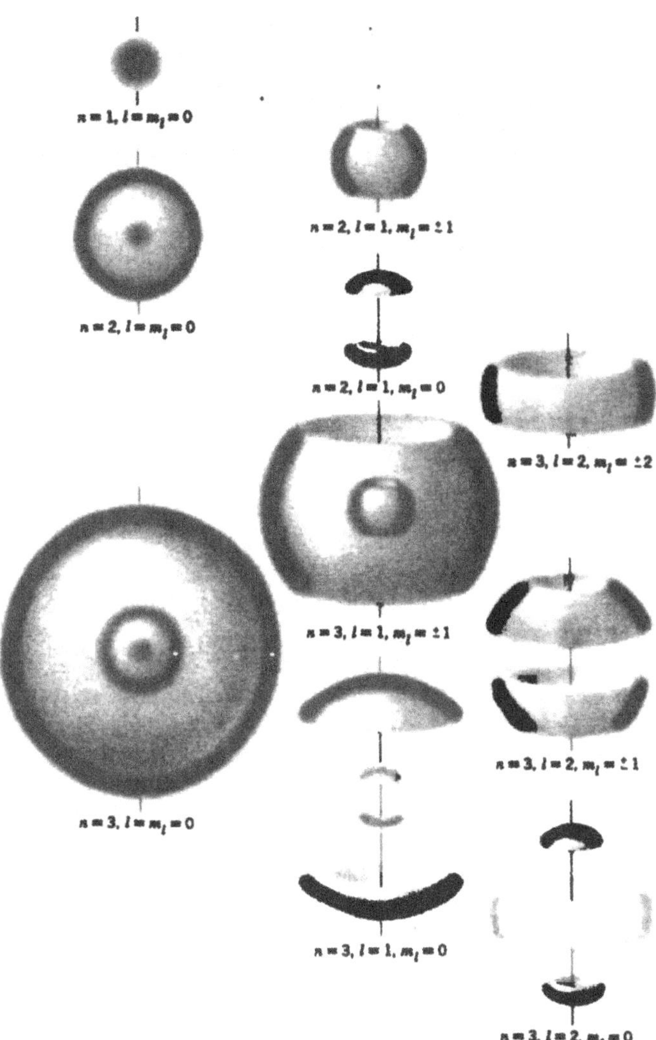

Fig. 3.3 An impression of the 'charge distributions' defined by some of the eigenfunctions of an electron in a hydrogen atom. Note that for a given value of n, addition of functions due to all values of l and m produces a spherically symmetrical distribution.(Reproduced with permission from Eisberg, R. and Resnick, R., *Quantum Physics;* published by Wiley, New York, 1985.)

demonstrated that no radiation occurs when an electron is in an eigenstate.

3.6 THE HYDROGEN ATOM

The model analysed above is obviously a model of the real-life hydrogen atom. Qualitatively, the result that the system is stable, with the atomic electron able to exist in one of a number of well-defined energy eigenstates, satisfies the experimental requirements. But a model must also make quantitative predictions.

First, (3.16) gives values for possible energy states of the electron. Intuition would suggest that the electron would be most likely to rest in the ground state where its energy is lowest. From (3.16), the actual ground state energy of the electron would be

$$E_g = -2.18 \times 10^{-18} \text{ J}.$$

More usefully, this can be expressed in the units of electron-volts (eV), where the value given is the electrical voltage required to accelerate an electron to the appropriate energy in joules. In these units

$$E_g = -13.6 \text{ eV}.$$

To see the importance of this value, refer back to the energy level diagram, Fig. 3.2. As n increases, the energy of the states increase until when n is very large, the energy is zero. Looking at the form of the eigenfunctions, this is seen to be the situation when the electron is, on average, so far away from the nucleus that it can be considered free of its attraction. This represents the conversion of the neutral atom into a positive ion and a free electron, and the energy required to do this is known as the ionisation energy of the atom. Theory has therefore predicted that the ionisation energy of hydrogen is 13.6 eV. This can be tested by experiments in hydrogen-filled discharge tubes and is found to agree well with experimental results.

Predictions can also be made about the size of the atom. The expression for the eigenfunction can give a value for the average distance of the electron from the nucleus.

For the ground state, therefore, it is possible to calculate $\langle r \rangle$, the mean distance of the electron from the nucleus, using the expression from Table 3.1. Care has to be taken, as the general expression (the three-dimensional version of (3.9)) gives the probability of the electron being in a volume element $dxdydz$ at position x. This is equivalent in spherical polar co-ordinates to the probability of the electron being in a volume element $r^2drd\theta d\phi$. What needs to be calculated in the present case is the probability $P'(r)$ of the electron being in a spherical shell of thickness dr at distance r from the nucleus. Thus

$$P'(r) = 4\pi r^2 \mid u_{1,0,0,}(r, \theta, \phi)|^2 \, dr. \qquad (3.18)$$

The value of $\langle r \rangle$ may now be calculated from the equation

$$\langle r \rangle = \int_0^\infty r \, P'(r) \, dr.$$

This gives

$$\langle r \rangle = 3h^2\varepsilon_o/(2\pi me^2) \quad \text{m}. \qquad (3.19)$$

The most probable distance of the electron from the nucleus, r_0, can also be calculated to be

$$r_0 = h^2\varepsilon_o/(\pi me^2) \quad \text{m}. \qquad (3.20)$$

This distance is normally known as the 'Bohr radius', in honour of Niels Bohr one of the founders of quantum physics. ($\langle r \rangle$ is greater than r_0 because $u(r)$ is not symmetric about r_0.)

Substituting, it is found that $\langle r \rangle = 8 \times 10^{-11}$ metres, a value in good agreement with the experimental values discussed in Chapter 2.

The model appears to accord well with experiment. One minor modification is needed before detailed quantitative comparisons can be made. In the original picture, it was assumed that the electron was moving in the field of an infinitely massive nucleus. In fact, the mass of the hydrogen nucleus is only some 1800 times that of the electron, so that both the electron and nucleus move around the centre of mass. However this may be taken care of by a minor adjustment, where the electron is seen as a particle with reduced mass μ rotating about the origin. The value of μ is given by the formula

$$\mu = m_e / (1 + m_e / m_n). \qquad (3.21)$$

Where m_e is the mass of the electron and m_n the mass of the nucleus. It can be seen that the correction is a relatively minor one.

The model appears satisfactory. However, to test it rigorously much more detailed experimentation is needed. The basis of such experimentation is discussed in the next chapter.

4

Atomic spectra

Without doubt, atomic spectroscopy is the most fruitful experimental technique in atomic physics.

4.1 SPECTROSCOPY AS A SOURCE OF INFORMATION ABOUT ATOMS

The physical principles underlying the technique of quantitative spectroscopy, or spectrometry were laid down by Kirchhoff and other workers in the 1860s, and experimentation throughout the second half of the nineteenth century produced the quantitative data needed for the unravelling of the structure of the atom. Most dramatically, line spectra from atomic discharges gave vital information.

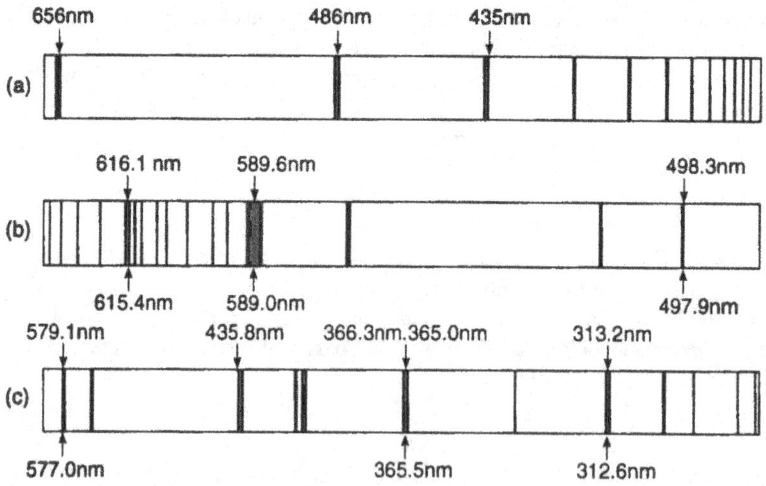

Fig. 4.1 Typical unretouched photographs of gas discharge spectra: (a) hydrogen, (b) sodium, (c) mercury. At this resolution many doublets and triplets cannot be resolved. (Reproduced from H.G Kuhn, *Atomic Spectra*; published by Longmans, London, 1969.)

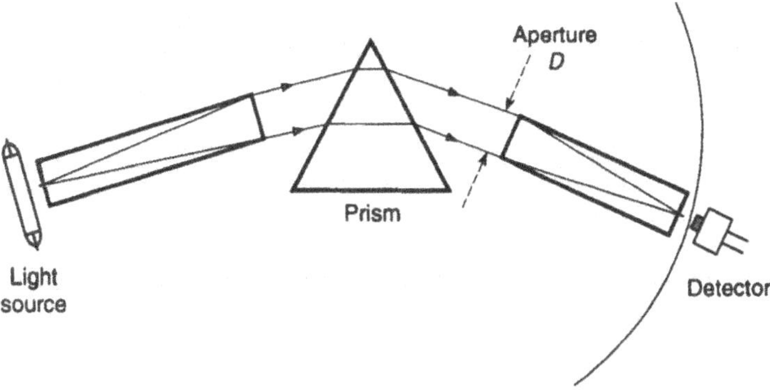

Fig. 4.2 Schematic diagram of typical optical spectrometer. The dispersion element is shown as a prism, but may be a diffraction grating.

Firstly, that each element possessed its own unique spectrum. Secondly, it was clear that each atomic spectrum consisted of a number of 'lines', or light of very well defined wavelength, and that those wavelengths appeared to have well-defined numerical relationships, characteristic of the particular atom (Fig. 4.1).

Spectroscopic measurements are now made over the whole electromagnetic spectrum from X-ray wavelengths to microwaves. There is no space here for a detailed discussion of spectroscopy, but the experimental investigations which laid the basis of atomic physics were made in the optical region (400–700 nm) and the techniques involved will be briefly described.

The heart of an optical spectrometer is some dispersion element, usually a prism or a diffraction grating. In essence, even the most sophisticated device is little different from the basic prism spectrometer shown in Fig. 4.2. Light from a source is directed onto the input slit. This light is focused into a parallel beam by the collimator and the refracted beam is focused again by the telescope onto the image slit. The tracking of the image slit produces a spectrum which may be observed by eye or recorded for quantitative analysis. Most of the classic experimental work was carried out by recording spectra on photographic emulsions (on film or glass plates) and a knowledge of the properties of the photographic emulsion was needed for detailed quantitative study of the intensity and structure of individual spectral lines. However, wavelength measurements may be made accurately by eye and are dependent only on the calibration of the spectrometer.

The setting up and calibration of a spectrometer is described in most textbooks of experimental physics. From the point of view of atomic spectroscopy, two parameters of a particular spectrometer are of importance.

First is the dispersion, which, for a prism spectrometer, depends on the variation of the refractive index of the prism material with wavelength. Second, and more important, is the resolving power of the spectrometer. Consider a source of light which gives out two perfect monochromatic waves at wavelengths λ and $\lambda + d\lambda$. If $d\lambda$ is large enough, the spectrometer output will be seen as two lines. If $d\lambda$ is decreased, the two lines will merge and will eventually be indistinguishable. The minimum value of $d\lambda$ which produces distinguishable lines at the output defines the resolving power of a given spectrometer.

In Fig. 4.2, the resolving power depends on the size of the prism aperture as well as on the prism dispersion. The prism and telescope together act like a slit and the spectral output is spread via Fraunhofer diffraction. Simple physical optics gives the result that, at a wavelength λ, the resolving power R, defined as $\lambda / d\lambda$, is given by

$$R - D\, d\Theta/d\lambda \tag{4.1}$$

where D is the prism aperture and $d\Theta/d\lambda$ is the dispersion of the prism material.

In a practical case, at a wavelength of 500 nm, a glass prism with a dispersion of 0.2 mradians/nanometre and an aperture of 50 mm, will have a resolving power of 10^4. It would be able to separate two lines at wavelengths of 500 nm and 500.05 nm.

Conversely, a perfect monochromatic line of wavelength 500 nm will be seen at the output slit as a Fraunhofer diffraction pattern of width approximately 0.05 nm. Thus the resolving power of a spectrometer also limits the precision with which the detail of a spectral line can be observed.

4.2 THE GENERAL CHARACTERISTICS OF GAS DISCHARGE SPECTRA

Gas discharge tubes such as the ubiquitous sodium streetlight are very familiar. But what makes them so useful in atomic spectroscopy?

The first necessity for studying atoms is that the atoms should be seen as individuals, rather than as an interacting collection, and a gas of atoms at relatively low pressure fulfils this requirement. Kinetic theory gives an expression for the mean free path of an atom in a gas

$$l \sim 1/4\pi r^2 N \quad \text{m}. \tag{4.2}$$

where r is the radius of an atom and N the number of atoms per unit volume.

For a tube of gas at room temperature and an easily achieved pressure of 0.001 torr (0.13 Pa), this gives a mean free path of around 0.25 m and a typical time between collisions with other atoms of 0.1 ms. Knowing the approximate size of an atom, it can be seen that each atom spends the vast majority of its time travelling freely, independent of interactions with other atoms. In other words, atoms in such a discharge tube can be considered independent of each other to a good approximation.

In order to give out light, the atoms must be excited in some way. The most obvious is by collision with free electrons that have been accelerated in an electric field. The gas discharge tube is essentially identical with Thomson's cathode ray tube of chapter 1, with the difference that the gas in the tube is now a necessity, not a nuisance. As shown in Fig. 4.3, electrons accelerated from the cathode accumulate kinetic energy and collide with atoms. No light is emitted within a few mm of the cathode (the Faraday dark space) and this strongly suggests that the electrons have to accumulate some minimum amount of energy before they can give it up to the atom. The atom absorbs the energy and this excitation energy is released again in the form of electromagnetic radiation. The study of spectra from gas discharges excited in this way is known as emission spectroscopy.

There is a separate technique known as absorption spectroscopy. In this case the gas is not directly excited, but is used to absorb radiation from a 'continuum' source such as a tungsten filament lamp, which emits light uniformly over the whole visible spectrum. The output observed is a continuous spectrum with black absorption lines. For a given gas, these lines are found to be at exactly the same wavelengths as the lines in the emission spectra, so the conclusion must be that energy from the source has been absorbed by the gas atoms.

Such absorption spectra were first observed in the solar spectrum by

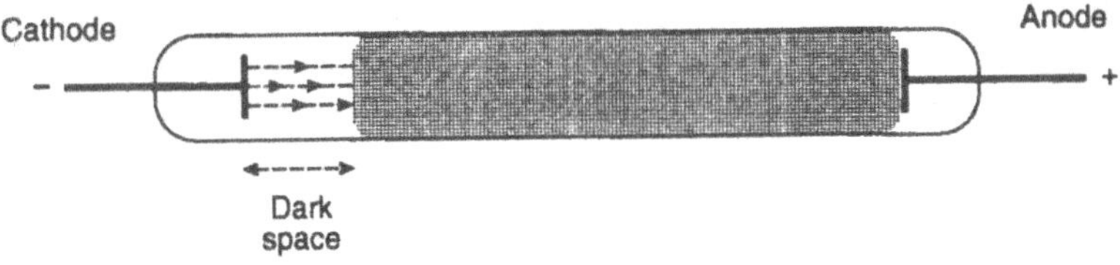

Fig. 4.3 Gas discharge tube. Electrons are accelerated from the cathode and collide with gas atoms which gain energy from the electrons then re-emit it in the form of light. Note the dark space where the electrons are not energetic enough to excite the atoms.

Fraunhofer in 1814. Light from the sun's central core is absorbed in the cooler outer corona. Later, the Fraunhofer lines were used to identify the major elements present in the sun and this finding gave a major impetus to the development of spectroscopy.

The above paragraphs give only the briefest outline of the techniques of atomic spectroscopy. Most particularly, they do not do justice to the difficulty of obtaining results and the skill and devotion of early spectroscopists. For example, it was very difficult to obtain pure samples of gas in a discharge tube, so the spectrum studied might well contain additions due to unwanted gases and impurities from the glass or quartz envelope and electrodes. Furthermore, electron impact may remove one or more electrons completely from the atom, so that spectra of ions may be present. Many sophisticated techniques were required to measure atomic spectra unambiguously.

Even in the most simple case, hydrogen gas normally exists in the form of diatomic molecules. Free electrons in the discharge rip some of these molecules apart and the observed spectrum consists of contributions from excited molecular hydrogen, atomic hydrogen and any other impurities in the tube. However, even with these complications, the study of the spectrum of atomic hydrogen gives vital information on atomic structure.

4.3 THE HYDROGEN SPECTRUM

The first view of the spectrum of atomic hydrogen may well be an anticlimax. Against a background of general mush, a series of three or (with care) four lines may be made out. But this spectrum can be considered as the starting point for the understanding of the physics of the atom.

The regularity of the visible hydrogen spectrum was first formulated mathematically by Balmer and this part of the spectrum now bears his name. Each line of the Balmer series is very narrow. In fact, using normal spectrometers, the observed linewidth is only limited by the resolution of the spectrometer. In other words, the radiation appears to be perfectly monochromatic.

Balmer showed that the wavelengths of the lines were described by the formula

$$\frac{1}{\lambda_n} = R\left(\frac{1}{4} - \frac{1}{n^2}\right)\ \mathrm{m}^{-1} \tag{4.3}$$

where λ_n is the measured wavelength of the line, n is an integer greater than 2 and R is a constant, known as the Rydberg constant after the physicist who first calculated its value.

Nowadays, the formula is more commonly written in terms of the frequency of the line

$$\nu_n = Rc\left(\frac{1}{4} - \frac{1}{n^2}\right) \text{ Hz.} \tag{4.4}$$

If the wavelength is measured in air, c is the velocity of light in air.

This formula agrees with experiment to within the precision of spectroscopic measurement.

Observation beyond the visible region showed other similar series, and the frequencies of the lines in the hydrogen spectrum can be generally defined by the expression

$$\nu_n = Rc\ (1/n_1^2 - 1/n_2^2)\ \text{ Hz.} \tag{4.5}$$

where n_1 defines the particular series and $n_2 > n_1$.

For example, taking $n_1 = 1$ gives the Lyman Series in the ultraviolet region and $n_1 = 3$ the Paschen Series in the infrared region. In each series, the equation defining the frequency of each spectral line appears as the difference of two terms, the 'fixed term' and the 'running term'. Because of this, spectroscopists referred to analysis in this form as 'term analysis'.

Comparison with the theory of the hydrogen atom in Chapter 3, shows that the term analysis corresponds precisely to the idea that the line spectrum is caused by transitions between the theoretical energy eigenstates. The fixed term corresponds to the energy of the lowest energy state. The running terms correspond to the energies of the states linked to the lowest state via the transition (whether emission or absorption). The transition between two states involves the absorption or emission of a quantum of electromagnetic radiation, with energy given by

$$\Delta E = h\nu \text{ J.} \tag{4.6}$$

where h is Planck's constant and ν the frequency of the particular line in the spectrum.

By comparison with equation (3.16), it can be seen that

$$R = \mu e^4/8c\varepsilon_0^2h^3 \quad \text{m}^{-1}. \tag{4.7}$$

where μ is the reduced mass of the electron, as defined in equation (3.21)

Substitution of the known values of fundamental constants once again gives

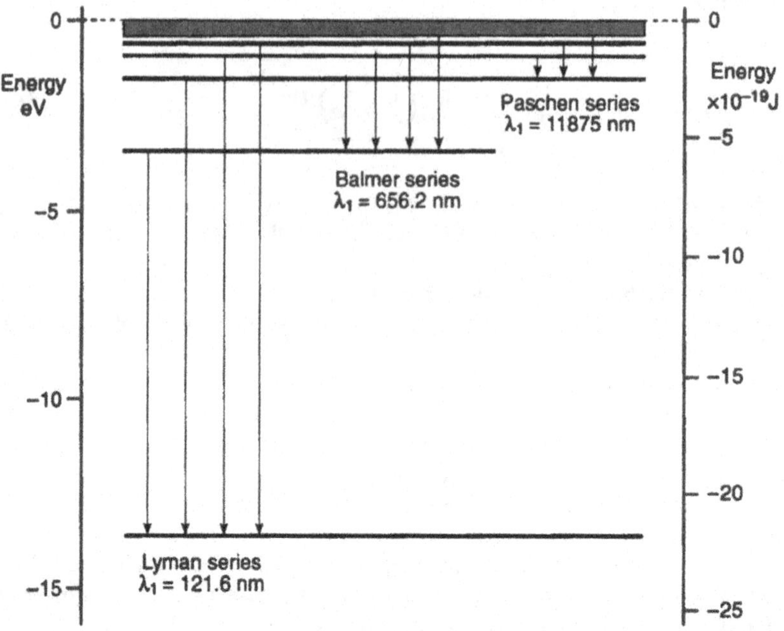

Fig 4.4 The three main series in the line spectrum of hydrogen. The Balmer series is in the visible region and is shown in Fig. 4.1(a).

agreement with experiment to within the precision of spectroscopic measurement, a triumphant vindication of theory.

The comparison of theory with experiment now makes it possible to state confidently that each spectral line defines a transition between two energy eigenlevels, where the energy eigenvalue is defined by the principal quantum number n, and the energy difference is given by

$$E_n - E_{n'} = \mathrm{h}\nu = \mathrm{h}c/\lambda \ \ \mathrm{J}. \tag{4.8}$$

The transitions can be drawn on an energy level diagram (first devised by Grotrian) as shown in Fig. 4.4. A transition to the value $n = \infty$ is equivalent to the complete removal of the electron from the field of the nucleus and hence the ionization of the atom.

The energies on the diagram are shown both in joules and in electron-volts. The use of the electron-volt is often easier for calculation, but more importantly, it gives a physical feeling for the processes involved in the discharge. For example, recall that the ground state of the electron is at an energy of -13.6 eV, so a free electron accelerated through the quite small voltage of 13.6 volts will possess enough energy to rip the bound electron away from the hydrogen nucleus.

4.4 HYDROGEN-LIKE SPECTRA

The atomic hydrogen spectrum agrees to a high degree of accuracy with the predictions of the quantum mechanical model. The spectra of similar systems agree to equally high precision. For example, singly-ionized helium consists of a single electron in the field of a nucleus of charge $2e$. As would be expected, its spectrum has series identical to those of hydrogen with the difference that the absolute values of frequencies are multiplied by a factor of 4. The same applies to doubly-ionized lithium, with a multiplication factor of 9. Such spectra can be observed, but they usually require more energetic excitation than that of a normal discharge tube. They were first observed in spark discharges in gas and are therefore traditionally called 'spark spectra'.

More interestingly, so-called alkali metals such as lithium, sodium and potassium also display absorption and emission spectral series like that of hydrogen (see Fig. 4.1). This strongly suggests that such atoms have some form of stable electron structure surrounding the nucleus, with a single loosely-bound outer electron interacting with external electrons or with radiation. This is the case, but a detailed discussion must be delayed until Chapter 8.

4.5 FINE STRUCTURE IN HYDROGEN AND HYDROGEN-LIKE SPECTRA

There is one further complication. Using high resolution spectroscopy, it is possible to resolve the various spectral lines into multiple lines and these multiple lines are referred to as the 'fine structure' of the spectral lines.

For example, the first line in the Balmer series is actually a doublet separated by a wavelength of 0.015 nm. Returning to the the quantum mechanical model, each value of the principal quantum number n provides a number of eigenstates with values of l as discussed in Section 3.5 and this can be displayed in an expanded Grotrian diagram, Fig. 4.5. The occurrence of fine structure shows that eigenstates of the same n but different l do not have exactly the same energy eigenvalue.

Another important phenomenon is also observed. It is found experimentally that not all transitions between eigenstates are possible. In fact, transitions only occur between states which differ in their values of l by ± 1 (Fig. 4.5).

In atoms other than hydrogen, broadening of the observed lines due to fine structure meant that the observed width of the lines in the series differed, depending on the fixed term and hence the lowest energy eigenstate in the transition. Thus spectral lines due to transitions that ended on a state where $l = 0$ tended to have relatively little broadening and were termed 'sharp'. Those that ended on a state where $l = 1$ were termed 'principal'. Those that ended on $l = 2$, 'diffuse' and so on.

This labelling is still in use today, so that in hydrogen a state where $n = 1$ and $l = 0$ is referred to a 1s state, one where $n = 2$ and $l = 0$, is a 2s state, one where $n = 2$ and $l = 1$ is a 2p state etc. (Fig. 4.5). The importance of this nomenclature will be seen in Chapter 7. The choice of labels is arbitrary, but, coincidentally, 's-states' have spherically symmetrical eigenfunctions.

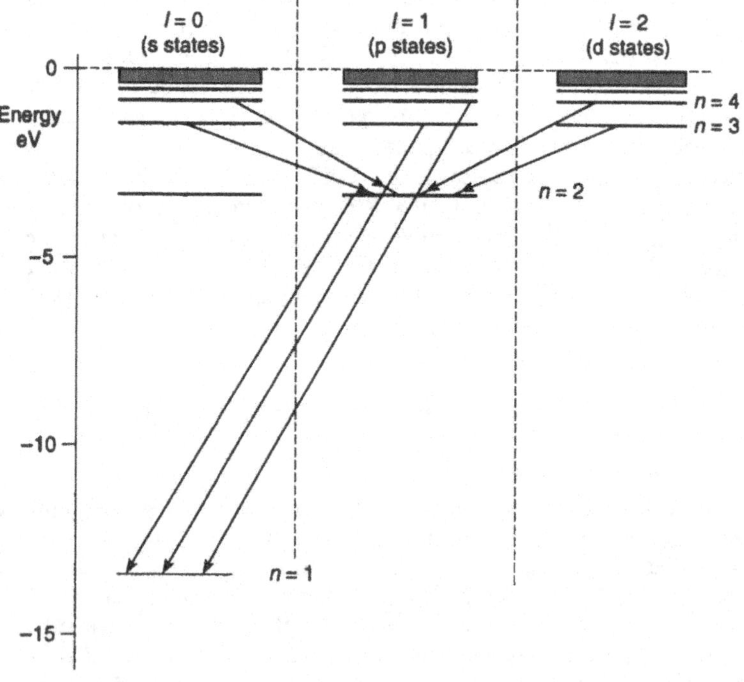

Fig 4.5 Expanded Grotrian diagram of energy levels in hydrogen. At this scale the energies for different values of l for a given value of n are identical.

5

Fine structure of spectral lines and electron spin

Although the major structure of the hydrogen spectrum is explained to a high level of precision by the quantum theory outlined in Sections 3.2 and 3.3, the reason for the fine structure of the spectral lines has still to be tackled.

5.1 FINE STRUCTURE IN THE HYDROGEN SPECTRUM

For example, examined under high resolution the 3s–2p line in the Balmer series turns out to be a doublet (Fig. 5.1). Referring back to (3.14), some extra term in the potential energy function V must exist which varies with the quantum number l and is therefore a function of the angular motion of the electron. This term must be small, as the effect it has on the energy eigenvalues is small.

The solution to this problem, suggested in 1925 by two young researchers

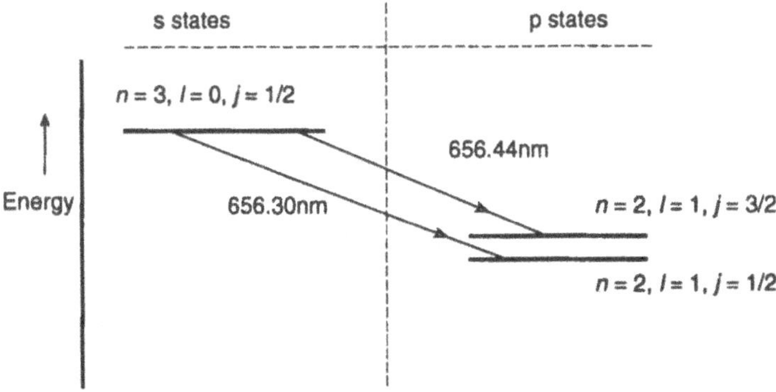

Fig. 5.1 Fine structure in the 3s – 2p transition in hydrogen. The energy spacing between the two p states has been greatly exaggerated (for explanation of labels, see Section 5.3).

Uhlenbeck and Goudsmit, appeared so strange that at first it was not generally accepted – that the electron possessed its own intrinsic spin.

If the electron is considered as a classical charged sphere, its energy in the field of the nucleus would be identical with that given in (3.14). However, if the charged sphere were spinning about its own axis, an extra energy term would occur.

Classically, a spinning ball of charge acts like a current loop and possesses a magnetic moment. An electron moving in an orbit about a nucleus appears like a moving charge in the field of a fixed charge but, from the frame of reference of the electron, it is the nucleus which appears to be moving. A moving charge produces a magnetic field, so the magnetic field due to the relative motion of the nucleus will interact with the magnetic moment of the spinning electron and the electron will gain a small extra energy. This interaction is therefore known as the 'spin–orbit interaction' and the energy due to the interaction is known as the 'spin–orbit coupling energy'.

5.2 CALCULATION OF THE SPIN–ORBIT COUPLING ENERGY

From classical electromagnetism, a negatively charged sphere spinning about its own axis possesses a magnetic moment $\boldsymbol{M}$, which is a function of the angular momentum of the sphere $\boldsymbol{s}$

$$\boldsymbol{M} = -K\boldsymbol{s}. \tag{5.1}$$

The form of the constant K depends on the distribution of charge on the sphere, but for a sphere of mass m with charge $-e$ uniformly distributed throughout its volume

$$K = e/2m. \tag{5.2}$$

If the sphere is moving with velocity υ in a uniform electric field $\boldsymbol{E}$, relativity theory gives the result that from the point of view of the sphere it 'sees' a magnetic field $\boldsymbol{B}$, where

$$\boldsymbol{B} = \boldsymbol{E} \times \upsilon / c^2 \quad \text{tesla.} \tag{5.3}$$

A body with magnetic moment $\boldsymbol{M}$ in a magnetic field possesses an energy

$$W = -\boldsymbol{M} \cdot \boldsymbol{B} \quad \text{J.}$$

Therefore from equations (5.1) and (5.3), a spinning charged body moving in an electric field will possess an energy due to its spin

$$W' = (K/c^2)\ \boldsymbol{s} \cdot \boldsymbol{E} \times \upsilon \quad \text{J.} \tag{5.4}$$

Considering the specific case of an electron moving at a distance $\boldsymbol{r}$ from a nucleus of charge $+e$,

$$\boldsymbol{E} = e\ \boldsymbol{r} / (4\pi\varepsilon_0\ |\boldsymbol{r}|^3) \quad \text{V/m.} \tag{5.5}$$

(Equation (5.5) is merely Coulomb's law, written so that the vector $\boldsymbol{r}$ is emphasized.)

Therefore equation (5.4) may be rewritten

$$W = (Ke/4\pi\varepsilon_0\ |\boldsymbol{r}|^3\ c^2)\ \boldsymbol{s} \cdot \boldsymbol{r} \times \upsilon \quad \text{J.} \tag{5.6}$$

Finally, the vector terms may be rearranged so that the angular momentum $\boldsymbol{l}$, which the particle possesses by virtue of moving in an orbit around the nucleus, is featured

$$W = (Ke/4\pi\varepsilon_0\ |\boldsymbol{r}|^3\ mc^2)\ \boldsymbol{l} \cdot \boldsymbol{s} \quad \text{J.} \tag{5.7}$$

This energy is the spin–orbit coupling energy or spin–orbit energy.

In fact, if the electron is moving around the nucleus and not parallel to it, a correction of ½ has to be made to (5.7). This is known as the 'Thomas precession' correction and gives a final value for the spin–orbit energy

$$W_{so} = (Ke/8\pi\varepsilon_0\ |\boldsymbol{r}|^3\ mc^2)\ \boldsymbol{l} \cdot \boldsymbol{s} \quad \text{J.} \tag{5.8}$$

Returning to the hydrogen atom, if it is postulated that the electron spins on its axis, (5.8) provides an extra energy term to add into the TISE, (3.14). However it is immediately obvious that the classical analysis does not give the correct result. Classically, W_{so} would vary continuously from a maximum value when the orbital and spin angular momenta were parallel, to a minimum when the momenta were in opposite directions. This would merely produce a broadening of the observed spectral line. It is necessary, therefore to replace the classical functions $\boldsymbol{l}$ and $\boldsymbol{s}$ with the equivalent quantum operators.

5.3 THE QUANTIZED ELECTRON SPIN

In order to introduce quantized theory, it is necessary to make one change to the classical expression (5.8). Since both the orbital and spin angular momenta are vectors, a total angular momentum vector $\boldsymbol{j}$ will exist, which will be the vector sum of $\boldsymbol{l}$ and $\boldsymbol{s}$. Using simple trigonometry, (5.8) may be rewritten

$$W_{so} = (Ke/16\pi\varepsilon_0|r|^3 mc^2)\,[\,|j|^2 - |l|^2 - |s|^2\,]\ \ J. \tag{5.9}$$

It is now possible to convert from classical angular momentum functions to quantized angular momentum operators. This is discussed in Appendix B and the upshot is the replacement of the values of $|j|^2$, $|l|^2$ and $|s|^2$ in (5.9) by the eigenvalues of the operators $|\mathbf{j}|^2$, $|\mathbf{l}|^2$ and $|\mathbf{s}|^2$. The result is

$$W_{so} = (Keh^2/64\pi^3\varepsilon_0\,|\mathbf{r}|^3 mc^2)\,[\,j(j+1) - l(l+1) - s(s+1)]\ \ J. \tag{5.10}$$

where j, l, and s are the appropriate quantum numbers for the eigenstate in question.

An important result is immediately obvious. For states where $l = 0$ (s-states). W_{so} must be zero. Now, since the spectral line due to the transition 3p – 2s (the first Balmer line) is a doublet, the only conclusion is that the 3p level must be split into two (Fig. 5.1). Therefore there must be only two possible values of W_{so} for the state where $l = 1$. However, for a given l, possible values of j are given by (Appendix B)

$$j = (l + s),\ (l + s - 1), \ldots, (l - s)$$

Therefore, if the spin orbit interaction is responsible for the splitting of this level, there can only be two values of j, and so the only possible value for s must be $^1/_2$. This analysis is confirmed by examination of the fine structure of other lines.

The important experimental result is obtained that the electron must have a spin angular momentum defined by the quantum number s which can only have the value $^1/_2$. The corollary, shown in Appendix B, is that the eigenvalue of the z-component of $\mathbf{s}$ can only have two values: $+^1/_2(h/2\pi)$ and $-^1/_2(h/2\pi)$. In other words, the electron can only have two spin states, 'spin up' and 'spin down' relative to some external co-ordinate system. This result is confirmed by other types of experiment, such as the Stern–Gerlach experiment (see Section 11.1).

In Chapter 3, three quantum numbers n, l, and m were introduced to define the state of the electron. To these must be added a fourth s. This defines the z component of the electron's spin and can have values +1/2 or -1/2.

There is one important final question. Looking back to (5.1) and (5.2), the value of K for a classical spherical electron would be $e/2m$. However it is found experimentally that $K = ge/2m$, where g has a value of 2.00232. This can only be explained via a quantum theory of fundamental particles.

5.4 OTHER CAUSES OF FINE STRUCTURE

One job remains, to add the expression for W_{so} to the potential energy term in (3.14). The solution of this equation would be complicated, but is simplified by a technique called time-independent perturbation theory (see Appendix C). This gives an approximate result which is valid, since W_{so} is very much smaller than the electric energy of the electron. The result is a correction, ΔE to the eigenvalue E_n

$$\Delta E = \int u_n{}^* \, W_{so} \, u_n \, dv \quad \text{J}. \tag{5.11}$$

where u_n is the eigenfunction appropriate to E_n, and the integration over dv signifies integration over all space.

Substitution from (5.10) and integration gives an absolute value for ΔE. However, this calculation is complicated and not immediately useful as there is another effect which gives a correction to E_n of the same order of magnitude.

The TISE, (3.10), assumes that the motion of the electron is so slow that there are no relativistic effects. However, this is not the case. Allowing for relativity gives an extra term in the kinetic energy part of (3.10) which is proportional to the fourth power of the momentum. Including this in the equation produces a further modification in the value of E_n which, unfortunately for experimentalists, is of the same order of magnitude as the effect of spin–orbit coupling. The situation produced by the two corrections to the simple TISE is summed up in

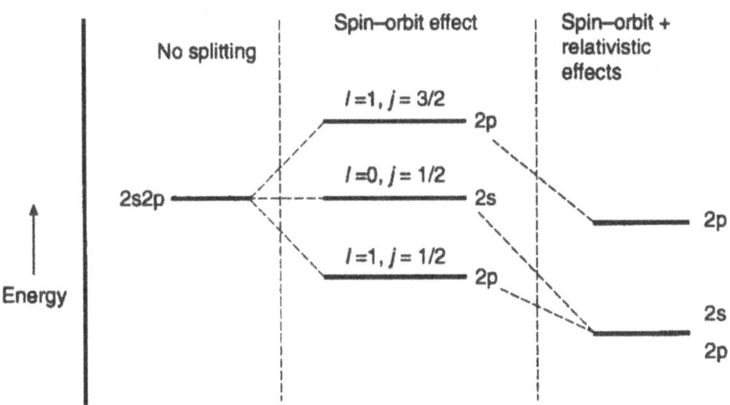

Fig 5.2 Contributions to the fine structure of the n = 2 energy level in hydrogen. Note that the total energy change including both spin-orbit coupling and relativistic effects is the same for the 2s and the 2p (j = 1/2) levels. This is not coincidental, but is a consequence of the solution of the relativistic version of Schrödinger's equation, the Dirac equation. However, quantum electrodynamic effects raise the 2s level slightly (an effect known as the Lamb shift).

Fig. 5.2.

In addition to the fine structure, an even smaller 'hyperfine splitting' can be observed. This is due to the fact that the nucleus itself has a magnetic dipole moment. Each electron energy state is split into two, due to the interaction of the electron dipole moment and the magnetic field due to the nuclear dipole. A calculation along the lines of that discussed in Section 5.2 gives an energy splitting of 9.4×10^{-25} J. Strictly, transitions between these two states are forbidden (see Section 6.6), but they do occur, giving radiation with a frequency of 1420 MHz. Radiation at this frequency, an unmistakable sign of the existence of hydrogen, plays an important part in the investigation of the distribution of hydrogen in interstellar space.

5.5 THE WIDTHS OF SPECTRAL LINES

Once the effects of fine structure are sorted out, the question of the actual width of a spectral line still remains. Using normal spectrometers, the measured linewidth is purely determined by the spectrometer resolution (see Section 4.1). However, it is reasonable to assume that the lines have some intrinsic width, since the alternative solution would mean a perfectly monochromatic wave produced by transitions between infinitesimally narrow energy levels.

Experimental work with instruments of much higher resolution enables the actual widths of spectral lines to be measured. However before they can give information about the widths of the energy levels, two sources of broadening, external to the atom have to be eliminated.

First is so-called 'Doppler Broadening'. Atoms in a discharge tube are moving

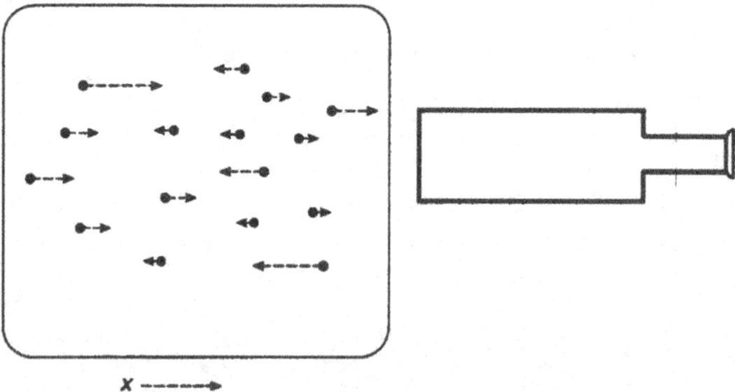

Fig. 5.3 Observation of radiation from moving atoms (x-components of velocities shown). Light from atoms moving towards the observer is Doppler shifted to higher frequencies, that from atoms moving away is shifted to lower frequencies.

relative to the observer. This means that the radiation observed from an individual atom is shifted in frequency: to higher frequencies if the atom is moving towards the observer, and to lower frequencies if it is moving away. The kinetic theory of gases gives an expression for the distribution of velocities of gas atoms in equilibrium at a given temperature. If the line is observed via a spectrometer collimator aligned along the x-axis (Fig. 5.3), the important question is the distribution of speeds along that axis. This is given by

$$P(\upsilon_x) \;\alpha\; \exp(-m\upsilon_x^2/2kT) \tag{5.12}$$

where υ_x is the speed in the x direction, m the mass of the atom, T the temperature of the gas and k is Boltzmann's constant.

A spectral line from a single atom with speed υ_x suffers a Doppler shift in its frequency

$$\Delta\nu/\nu = \upsilon_x/c. \tag{5.13}$$

Therefore the observed spectral line, due to the sum of light from a large number of atoms moving with velocities defined by (5.12) will be broadened and the broadening is found from (5.12) and (5.13) to be

$$I(\nu) = I(\nu_0)\exp -[c^2 m(\nu - \nu_0)^2/2kT_0^2] \tag{5.14}$$

where ν_0 is the central frequency, equivalent to radiation from those atoms with $\upsilon_x = 0$.

This 'Doppler broadening' masks any other sources of broadening. For a gas discharge temperature of 500K, the broadening, defined as the frequency spread at half height of the curve, is of the order of 10^9 Hz.

One further source of broadening external to the atom is 'pressure broadening'. This is due to collisions between atoms. Simply, if the mean time between collisions is τ_c, the broadening of the line is of the order of $1/\tau_c$ s^{-1}. This effect is noticable in radiation from gas discharges at moderate pressures, but may be eliminated by decreasing the gas pressure until τ_c is longer than τ, the lifetime of the state.

Sophisticated experimental techniques can eliminate effects due to Doppler and pressure broadening. The remaining line broadening, known as 'natural broadening' is due to processes internal to the atom.

The shape of the naturally broadened spectral line can be described by the Lorentz function

$$I(\nu) = I(\nu_o)\,\gamma^2/[\gamma^2 + (\nu - \nu_0)^2]. \tag{5.15}$$

The term γ is known as the natural linewidth of the spectral line. Typically, it is of the order of 10^8 Hz.

The various types of broadening of spectral lines and their relative magnitudes are summed up in Fig. 5.4.

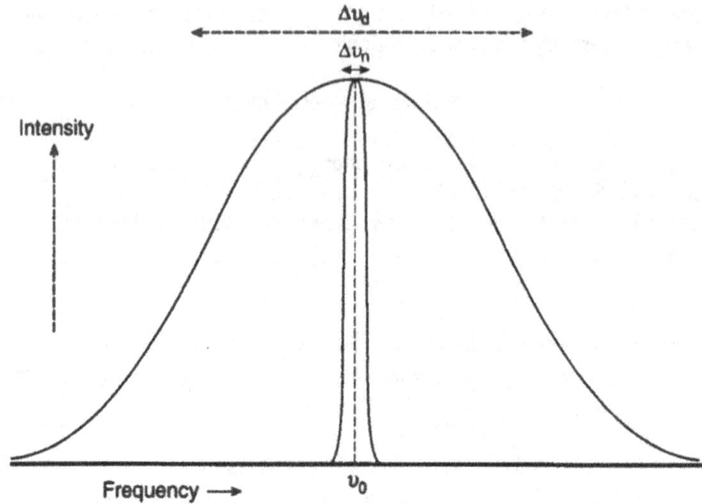

Fig. 5.4 Broadening of a spectral line in a gas discharge. The Doppler broadening Δv_d, defined by (5.14) is typically much larger than the natural broadening Δv_n, given by (5.15). Pressure broadening, by definition, depends on the pressure in the discharge gas. For typical discharge tubes, it is rather larger than natural broadening, but smaller than Doppler broadening.

6

The interaction of atoms and radiation

The last two chapters have demonstrated the success of the quantum mechanical theory of the simplest atom, but one further area must be explored before extending this theory to more complex atoms. The interaction of atoms with electromagnetic radiation is central to experimental atomic physics, so it is appropriate that a chapter on this topic is at the centre of this book.

6.1 ABSORPTION AND EMISSION OF RADIATION

Interestingly, the basic phenomena of interaction between atoms and e.m. fields can be discussed in some detail without the need of a detailed model of the atom. The theory was laid down by Einstein in the early 1900s and, to a great extent, lay fallow for the next fifty years until the invention of lasers.

The model of the atom requires only that the atom has two stationary energy states, 1 and 2, with energies E_1 and E_2 ($E_1 < E_2$). Transitions can take place between the two states and each transition will involve the absorption or emission by the atom of electromagnetic radiation of frequency ν, where the value of ν is given by

$$E_2 - E_1 = h\nu \quad \mathrm{J}. \tag{6.1}$$

This picture was postulated by Einstein from a consideration of the experimental analysis of line spectra and from Planck's work on black body radiation a year or so earlier. In that work, Planck had been forced to conclude that the radiant energy in a field at a particular frequency ν had to be quantized, meaning that energy could only be absorbed or emitted in multiples of a finite quantum of energy, which had the value $h\nu$ joules. The constant h was given Planck's name.

Given this picture, Einstein postulated that only three types of interaction process are possible between the atom and the field (Fig. 6.1).

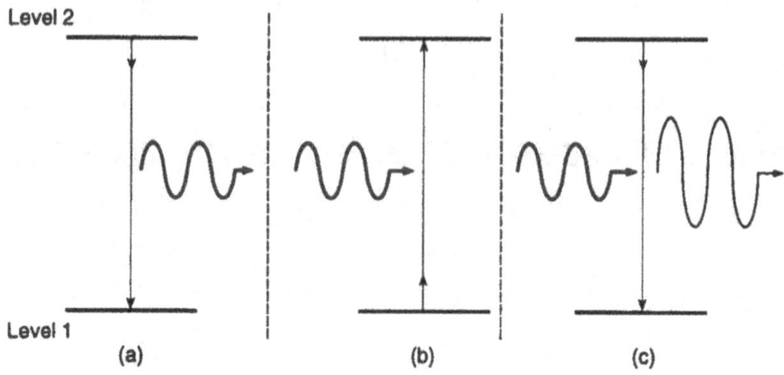

Fig. 6.1 Types of interaction between a two-level atom and an electromagnetic wave. (a) An atom in level 2 spontaneously emits electromagnetic radiation. (b) The atom absorbs energy from the wave and is excited from level 1 to level 2. (c) The atom is stimulated to fall from level 2 to level 1 and add energy to the wave.

First is spontaneous emission, where an atom in state 2 spontaneously emits a quantum of radiation to the field and falls to state 1. Assuming that the probability of this occurring is independent of time, it is easy to write the probability of spontaneous emission in the short time dt, dW as

$$dW = A_{21}\, dt \tag{6.2}$$

where A_{21} is a constant coefficient.

The second process is absorption, where an atom in state 1 absorbs a quantum of energy from the field and is excited to state 2. In this case, it is reasonable to postulate the probability of absorption as being proportional to the energy density of the field at frequency ν, $\rho(\nu)$. Then the probability of absorption dW' is given by

$$dW' = B_{12}\, \rho(\nu)\, dt \tag{6.3}$$

where B_{12} is a constant coefficient.

Third, and symmetric with the process of absorption, is the process of stimulated emission. In this case, an atom in state 2 interacts with the field by emitting a quantum of radiation and falling to state 1. As before, the probability of this occurrence, dW'' is given by

$$dW'' = B_{21}\, \rho(\nu)\, dt. \tag{6.4}$$

It must be noted that there is nothing new in the twin concepts of absorption and stimulated emission. They occur in classical interactions between fields and oscillating systems. A simple analogy is of pushing a garden swing. When the pusher pushes the swing in phase, energy is transferred from the pusher to the swing. Out of phase, energy is transmitted from the swing to the pusher – possibly with painful results. A conclusion from this symmetry is that $B_{12} = B_{21}$.

Ironically, as will be seen later, the most difficult process to analyse is spontaneous emission – the most common phenomenon.

6.2 EQUILIBRIUM BETWEEN ATOMS AND THE RADIATION FIELD

From the definitions in (6.2), (6.3) and (6.4), it is now possible to discuss the situation when a gas of atoms is in equilibrium with a radiation field in a closed volume. Assuming N_1 atoms exist in state 1 and N_2 in state 2, equations can be written down defining the rate of population and depopulation of the two levels. These are known as 'rate equations'.

For atoms in state 1

$$dN_1/dt = -B_{12} N_1 \rho(\nu) + N_2 (B_{12} \rho(\nu) + A_{21}) \qquad (6.5)$$

and for atoms in state 2

$$dN_2/dt = +B_{12} N_1 \rho(\nu) - N_2 (B_{12} \rho(\nu) + A_{21}). \qquad (6.6)$$

At equilibrium $dN_1/dt = dN_2/dt = 0$, so that (6.5) or (6.6) can be reorganized to give an equilibrium value for N_2 / N_1

$$N_2 / N_1 = B_{12} \rho(\nu) / (B_{12} \rho(\nu) + A_{21}). \qquad (6.7)$$

Now, if both the system of atoms and the electromagnetic field are in equilibrium, the value of N_2 / N_1 must be given by Boltzmann's law

$$N_2 / N_1 = \exp[-(E_2 - E_1)/kT] \qquad (6.8)$$

and $\rho(\nu)$ by Planck's law

$$\rho(\nu) = 8\pi h\nu^3/c^3[\exp(h\nu/kT) - 1] \quad \text{Js/m}^3. \qquad (6.9)$$

From (6.7), (6.8) and (6.9) a relation can be worked out between A_{21} and B_{12} :

$$A_{21} / B_{12} = 8\pi h\nu^3/c^3 \quad \text{Js/m}^3. \qquad (6.10)$$

This surprisingly simple relationship will be very useful later.

6.3 THE PHYSICS OF THE *A* AND *B* COEFFICIENTS

Some very interesting inferences may be drawn from Einstein's picture of the interaction process. The basic assumption is that energy must be absorbed and emitted by the atom in the form of quanta (6.1). This hypothesis was confirmed by Lenard's experiment, where the energy of photoelectrons ejected from a cathode by incident light was found to be a function of the frequency of the light. In the famous formula

$$E_{el} = h\nu - \phi \quad \mathrm{J} \tag{6.11}$$

where E_{el} is the kinetic energy of the electron and ϕ is the amount of energy required to remove the electron from the cathode, the 'work function'.

As will be seen shortly, (6.11) may be derived in a completely satisfactory way from the interaction of a quantum mechanical atom with a classical electromagnetic field. The common statement that it provides unequivocal evidence for the quantum nature of the electromagnetic field is unequivocally wrong.

However, when spontaneous emission is considered, the situation is very different. Equation (6.11) is, in fact, a statement of conservation of energy, but in any interaction, not only energy but momentum must be conserved. It is this fact which produces unequivocal evidence for quantization of the electromagnetic field. Einstein pointed out that if an amount of energy $h\nu$ were transferred in an absorption or emission process, an amount of momentum $h\nu/c$ would also have to be transferred. Conservation of momentum dictates that any absorption or emission process has to be directional. This presents no problem for the processes of absorption and stimulated emission. For stimulated emission, for example, the radiation emitted by the atom has to be in the direction of and in phase with the stimulating radiation.

However, the spontaneous emission process also has to be directional, something not consistent with classical theory. It is this fact that makes it necessary to consider the quantization of the electromagnetic field, a subject that will be discussed later in this chapter.

The next step is the calculation of the Einstein coefficients for the nuclear atom. The easiest way to do this is first to calculate the absorption/stimulated emission coefficient, *B*. To do this, a model of the interaction of field and atom must be developed.

6.4 PHYSICAL MODEL OF THE FIELD–ATOM INTERACTION

It is useful to start with a classical picture of the interaction. Consider the electron as a symmetrical cloud of negative charge surrounding the massive nucleus. An external electromagnetic field with electric field vector E_x in the x-direction will distort this charge and produce an electric dipole with dipole moment

$$D = \langle x\mathrm{d}q \rangle \quad \mathrm{Cm}$$

where dq is the charge density at distance x from the origin (Fig. 6.2).

The value of D will depend on the force between the negative cloud and the positive nucleus. If E_x is a sinusoidal function of time, D will also oscillate. If, by any chance, the attractive force between the cloud and nucleus is such that the system has a resonant frequency ν_D, the interaction between the field and the atom will be maximum when the frequency of oscillation of the field is equal to ν_D. If the field is in phase with the oscillations of the charge cloud, the atom will gain energy from the field. If it is 180° out of phase, the field will gain energy from the atom.

Converting from the classical picture to the quantum mechanical model involves a number of steps. As before, only two eigenstates are considered, with eigenfunctions u_1 and u_2 and energy eigenvalues E_1 and E_2, where $E_2 - E_1 = h\nu_{12}$ (note that these subscripts are only labels specifying the states and do not refer to quantum numbers).

Second, it is assumed that the force on the electron due to the external electromagnetic field is very small compared with the force due to the nucleus.

Fig. 6.2 Production of electric dipole by distortion. The vaue of the dipole moment is given by $\langle x\mathrm{d}q \rangle$. For a hydrogen atom, this is equivalent to $e\langle x \rangle$

In this case the external field only introduces a small perturbation on the solutions of the Schrödinger equation. However, in this case the perturbation is time-dependent, so that rather than using the time independent eigenfunctions u_1 and u_2, the full eigenfunctions Ψ_1 and Ψ_2 must be used. These eigenfunctions are $\Psi_1 = u_1 \exp[-i(2\pi E_1 t/h)]$ and $\Psi_2 = u_2 \exp[-i(2\pi E_2 t/h)]$.

The quantum mechanical analogy to the distortion of the classical charge cloud is given by the writing down of a general state function for the system, Ψ. This function is a linear combination of the unperturbed eigenfunctions, with coefficients which define the proportion of each eigenfunction in the total function. These coefficients therefore define the probability of finding the atom in a particular eigenstate and may be time-dependent. Thus

$$\Psi = c_1(t)\,\Psi_1 + c_2(t)\,\Psi_2$$

or

$$\Psi = c_1(t)\,u_1 \exp[-i(2\pi E_1 t/h)] + c_2(t)\,u_2 \exp[-i(2\pi E_2 t/h)]. \qquad (6.12)$$

To illustrate, if at time $t = 0$, $c_1(0) = 1$ and $c_2(0) = 0$ the system would be in state 1. If after time T, $c_1(T) = 0$ and $c_2(T) = 1$, the system would now be in state 2. Sometime within that period the system would have undergone a transition from state 1 to state 2.

The quantum mechanical analogue of the classical dipole moment is:

$$D = -\int_{-\infty}^{\infty} ex\Psi^*\Psi \, dv \quad \text{Cm} \qquad (6.13)$$

where the integration is over all space.

As in the classical case, the size of this dipole moment defines the magnitude of the interaction. It can be seen immediately that if Ψ is an unperturbed eigenstate of the system, D is zero. This is because eigenstates all have symmetry about the origin so that the integral in (6.13) is identically zero. Hence, as stated in Chapter 3, an electron in an eigenstate does not radiate.

The next job is to calculate c_1 and c_2. To do this, it is necessary to return to the full time-dependent Schrödinger equation, adding the extra perturbation energy due to the effect of the electric field on the electron. If the e.m. field is assumed to be monochromatic with frequency ν_f, the extra energy term will be

$$E = e\,x\,E_x \cos(2\pi\,\nu_f\,t) \quad \text{J}. \qquad (6.14)$$

The equation must then be solved. The calculation proceeds by the technique of time-dependent perturbation theory to produce first-order approximate solutions for c_1 and c_2 in the case when $\nu_f \sim \nu_{12}$. The details of the calculation are outlined in Appendix D.

For an electron which starts in state 1, the approximate solution of interest is the probability of transition to state 2. This is given by the probability of finding the electron in state 2 at time t

$$P_2(t) = |c_2|^2 = (4\pi^2/h^2)|D_{12}|^2\, t^2\, (\sin\xi/\xi)^2 \tag{6.15}$$

where $$\xi = 2\pi\,(\nu_{12} - \nu_f)\, t/2$$

and $$D_{12} = \int_{-\infty}^{\infty} ex\, u_2^*\, u_1\, dv$$

The term D_{12}, usually known (for reasons which need not concern us) as the 'quantum dipole moment matrix element', defines the strength of the interaction, just as it did in the classical case.

The physical meaning of (6.15) is not very transparent, so must be discussed in some detail.

6.5 FERMI'S 'GOLDEN RULE' FOR TRANSITIONS

Equation (6.15) appears a little odd. Firstly, the probability of transition appears to vary as the square of time. Secondly, and more importantly, the bracket term $(\sin\xi/\xi)$ becomes indeterminate at the 'resonance frequency' where $\nu_{12} = \nu_f$. These peculiarities are due to the fact that the calculation has dealt with the interaction of a perfectly monochromatic wave with an atom which possesses infinitely narrow energy levels. If either or both of these restrictions are removed, a physically comprehensible result is obtained.

First, consider the atom interacting with an electromagnetic field which is not monochromatic but has a constant energy density $\rho(\nu)$ at frequencies near to ν_{12}. It is then necessary to rewrite (6.15) in terms of an integral over all frequencies (noting that only those near ν_{12} will be important)

$$P_2(t) = |c_2|^2 = (4\pi^2/h^2)\,|D_{12}|^2 \int t^2\, \rho(\nu)\,(\sin\xi/\xi)^2\, d\rho \tag{6.16}$$

Changing variables and carrying out the integration leads to a more straightforward expression for $P_2(t)$, and hence for the transition rate $dP_2(t)/dt$

$$dP_2(t)/dt = (4\pi^2/h^2)\, |D_{12}|^2\, \rho(\nu_{12}). \tag{6.17}$$

This defines the transition rate for an atom in state 1 to be excited into state 2 and vice versa. It is normally known as 'Fermi's golden rule'. It should be noted that the result is only a first-order approximation, but is quite good enough for most cases.

By comparison with section 6.1, it can be seen that the Einstein B coefficient is given by

$$B_{12} = (4\pi^2/h^2)|D_{12}|^2. \tag{6.18}$$

This makes it possible to calculate a physical value for Einstein's B.

The non-physical nature of (6.15) can also be removed in a different way. If the field is assumed monochromatic, but the atom is excited from an infinitely narrow state 1 to a state 2 which is part of a continuum of states, a similar calculation may be made. In this case

$$P_2(t) = |c_2|^2 = (4\pi^2/h^2)|D_{12}|^2 \int t^2 \rho(E)(\sin\xi/\xi)^2 \mathrm{d}E \tag{6.19}$$

where $\rho(E)$ is the density of states at energies near to E_2.

In this case, changing variables and integrating produces the result

$$\mathrm{d}P_2(t)/\mathrm{d}t = (4\pi^2/h)\,|D_{12}|^2\,\rho(E). \tag{6.20}$$

Equation (6.20) defines a physical situation where an electron is removed from a bound state to free space. It therefore describes the situation which takes place in the photoelectric effect. Inspection will show that this result is consistent with the photoelectric equation (6.11), even though it has been derived by considering a perturbation due to a classical electromagnetic field.

6.6 ALLOWED AND FORBIDDEN TRANSITIONS

Equation (6.19) shows that the probability of a transition between states 1 and 2 depends on the dipole moment matrix element D_{12}

$$D_{12} = \int_{-\infty}^{\infty} ex\, u_2^*\, u_1\, \mathrm{d}v. \tag{6.21}$$

This depends critically on the form of the eigenfunctions u_1 and u_2. Referring back to (3.17) and converting to an integration over all space in spherical polar co-ordinates, the expression for D_{12} becomes

$$D_{12} = e\iiint r^2 dr\, d\theta\, d\phi\, R^*_{n',l'}(r)\, Y^*_{l',m'}(\theta,\phi)\,.\, x\,.\, R_{n,l}(r)\, Y_{l,m}(\theta,\phi). \qquad (6.22)$$

The quantum numbers (n',l',m') are those defining state 2 while (n,l,m) are those appropriate to state 1. To carry out the integral, of course, the Cartesian direction x defining the direction of the electric field vector must be rewritten in terms of polar co-ordinates

$$x = r \sin\theta \cos\phi. \qquad (6.23)$$

The value of D_{12} for the transition between any two states can be worked out by substituting the appropriate expressions (say from Table 3.1) and carrying out the integration. This is a wearisome process but fortunately some very interesting conclusions can be deduced without carrying out the whole integration.

First, consider the integration over ϕ. When multiplied out, terms in ϕ will consist of products like

$$\exp[-i(m'\phi)]\exp[+i(m\phi)]\exp[\pm\, i\phi]$$

where the first term comes from state 2 the second from state 1 and the third from the polar expression for x, (6.23). By rewriting this, we can see that the integration over ϕ will consist of a sum of terms like

$$\int_0^{2\pi} \exp\,[i(m - m' \pm 1)\phi]d\phi.$$

But an integral over such a term is identically zero unless the exponent is zero, in which case the integral is π. This leads to the important conclusion that

$$D_{12} = 0 \text{ unless } m - m' = \pm 1.$$

Thus no transition will take place unless the m quantum numbers of the two states differ by $\pm$ 1. This is known as a selection rule for transitions.

A similar (but more complicated) consideration of the θ-dependent part of the equation gives rise to the further selection rule

$$D_{12} = 0 \text{ unless } l - l' = \pm 1. \qquad (6.24)$$

This can be understood qualitatively by considering the dipole moment produced by a combination of 'charge clouds' due to different eigenstates (Fig. 6.3). Symmetry considerations show that D_{12} is zero unless $l' \neq l$. In practical terms, this result means that there will be no 3s - 2s transitions in the hydrogen spectrum, something already observed experimentally.

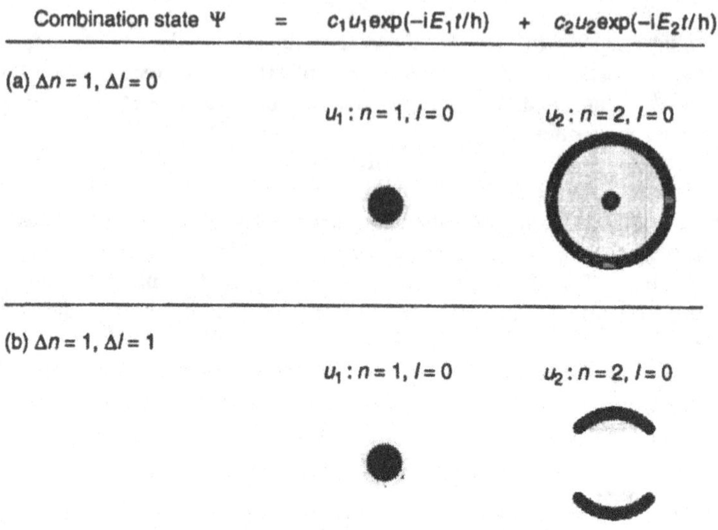

Fig. 6.3 Schematic picture to illustrate forbidden and allowed transitions. In (a) there is no equivalent dipole moment hence no interaction with the field and no transition. In (b) there is an equivalent dipole moment with resonant frequency $(E_2\text{-}E_1)/h$ Hz.

Consideration of the r-dependent part of the integral does not provide any further selection rule, so that (for example) a 3s – 3p transition is allowed and can be observed.

Working out the absolute magnitude of D_{12} requires a complete calculation of the integral (for most common transitions, the value is of the order of $-er_0$, where r_0 is the Bohr radius defined in Chapter 3). The analysis has now given a physical basis, first for the calculation of the Einstein B coefficient, and second for the experimental fact that transitions only take place between certain eigenstates.

Finally, it should be noted that 'forbidden' transitions are not completely forbidden. The analysis has been based on the idea of an external field inducing a dipole moment in the electron 'cloud of charge'. Even if such a dipole moment is not induced, a 'quadrupole moment' or moment of higher order may be induced. However, the interactions in these cases are very much smaller than those involved in 'dipole' transitions and so the probability of their occurrence is very much smaller. Thus, although spectral lines due to 'forbidden' transitions are observed, they are very much weaker than those due to 'allowed' transitions.

6.7 SPONTANEOUS EMISSION

The last subject to be approached is the spontaneous emission of radiation. In fact, having calculated the B coefficient for a given transition, it is easy to work out the equivalent value of A from (6.10). This gives a value for the transition rate for spontaneous emission. For common transitions, A is of the order of $10^8\,s^{-1}$. Thus, in an ordinary discharge tube where stimulated emission can be neglected, the mean lifetime of an atom in an excited state can be taken to be of the order of A^{-1} s.

This also enables an estimate of the natural linewidth of the light emitted by the transition. Taking Δt as a measure of the uncertainty in the time the electron spends in the excited state, the Heisenberg uncertainty principle gives an estimate of the uncertainty in the energy which the electron possesses in that state, ΔE

$$\Delta E \Delta t \approx h/2\pi. \tag{6.25}$$

This may be thought of in terms of a 'spread' in the energy of the state, so that there will be an equivalent spread in the frequency of the radiation emitted

$$\Delta \nu \approx \Delta E/h \approx A/2\pi \quad s^{-1}. \tag{6.26}$$

This makes the natural broadening of a spectral line to be of the order of 10^7–10^8 Hz, a value confirmed by experiment.

However satisfactory this result may be, there is still a major problem. If the electron is in an excited state in the absence of any electromagnetic field, there is no perturbing force and hence no reason for transition. There must be some further effect, not yet discussed.

6.8 SPONTANEOUS EMISSION AND THE QUANTIZATION OF THE E.M. FIELD

Absorption and stimulated emission of radiation can be explained satisfactorily by the theory described above, where a classical electromagnetic wave provides a time-dependent perturbation to the energy of the atomic electrons. Because the analysis involves the interaction of a classical electromagnetic field with a quantized atom, it is commonly described as 'semi-classical'. However, semi-classical theories cannot describe in any physically satisfactory way the process of spontaneous emission, since, by definition, when an atom emits spontaneously there is no pre-existing external field and hence no perturbation of the atom's eigenfunctions.

In fact, in his original, pre-quantum mechanical, work Einstein had showed that classical electromagnetic field theory was deficient in its explanation of the interaction between radiation and quantized atoms. This followed from his analysis of the fact that such an interaction must not merely conserve energy but also momentum. For stimulated emission, this provided no difficulty, since it simply implied that the emitted radiation had to be in the same direction as, and in phase with, the stimulating radiation. Similarly, in absorption, momentum was transferred from the field to the atom. However, for spontaneous emission, a detailed analysis of the momentum balance in the atom–field system led to the inescapable conclusion that spontaneous emission had to be a directional process. This could not be made consistent with classical theory.

The corollary is straightforward. Just as it was necessary when dealing with dynamics to convert from classical dynamical variables to quantized dynamic operators, it is necessary to convert from classical e.m. field variables to quantized field operators. This is not a simple, unambiguous process and the method used requires considerable theoretical justification. However, the theory of quantized electromagnetic fields (or quantum electrodynamics as the most general form of the theory is known) is so important to modern developments that its basis must be sketched out. An introduction to the mathematical analysis is given in Appendix E

The starting point for the transformation is the link between classical electromagnetism and classical mechanics, the fact that an electromagnetic field carries energy and so a Hamiltonian (energy) function H can be written down for the field. By comparison with the analysis outlined in Chapter 3, quantization involves the conversion of this function to an equivalent Hamiltonian operator, H. This is accomplished by considering the expression for the energy of a single mode of a classical e.m. field. It is found that this expression takes the same functional form as the Hamiltonian for a simple harmonic oscillator. In the case of the oscillator the variables are the position and momentum functions. The variables for the field are the $\boldsymbol{E}$ and $\boldsymbol{B}$ vectors, but the functional relationship is identical.

The quantized result for the harmonic oscillator was derived from the classical one by replacing the classical variables p and x by quantum operators $\mathbf{p}$ and $\mathbf{x}$. Similarly, quantization of the e.m. field may be achieved by replacing the equivalent variables by operators. However, as noted above, this is not completely straightforward and unambiguous.

The major problem is the problem of 'observability'. One of the principles of quantum mechanics is the fact that only certain types of operators (known as Hermitian operators) can be taken as equivalent to the physical process of making a measurement. The operators $\mathbf{x}$ and $\mathbf{p}$ (defined in Chapter 3) are of this type, so that quantum measurements of position and momentum can be compared directly with their classical equivalents. However, the quantum operators equivalent to **E** and **B**, the classical electric and magnetic field amplitudes, are not Hermitian and therefore cannot represent observable quantities. Physically, this is fairly easily understood. Radiation detectors do not measure electric or magnetic fields, they detect energy falling on their photosensitive surfaces. Therefore, in fact, one observable for an electromagnetic field is the energy in the field. Following through the quantization process it is found that the energy operator is Hermitian. For a single mode of the field, the eigenstates E_n of this energy operator are given by the simple equation

$$E_n = (n + ½)\ \mathrm{h}\ \nu_m \tag{6.27}$$

where n is an integer, h is Planck's constant, and ν_m the frequency of the mode. The integer n defines the number of quanta of energy in the mode so the operator is usually known as the 'photon number operator'.

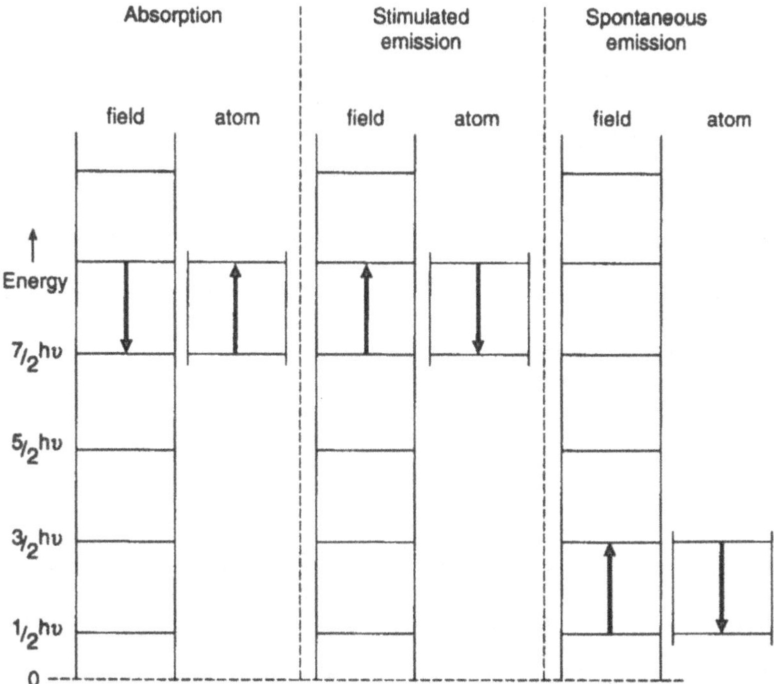

Fig. 6.4 Diagram showing the three possible types of interaction between the quantized energy states of an atom and of an electromagnetic field. In spontaneous emission, the atom interacts with the ground state of the field, which has zero point energy ½hν.

As in the case of **x** and **p**, there is another operator paired with the photon number operator, connected to it via the Heisenberg uncertainty principle. This is known as the 'photon phase operator'. The connection between these two operators may be seen simply by considering the experimental operation of photography. In classical optics, the surface of a photographic plate registers precisely the amount of energy falling on it, but stores no information on the relative phases of waves arriving from different objects, giving rise to the 'flat' photographic image. Conversely, it might be possible to devise some sort of interferometer which measured the relative phases, while giving no information about intensities. However, in the quantum case the situation is quite complicated.

It is now possible to understand the reason for the existence of spontaneous emission without any further mathematical analysis. If (6.27) defines possible states of the radiation field, a much tidier picture of interaction between field and atom can be drawn (Fig. 6.4). Absorption involves a single mode of the field losing one quantum of energy while the atom gains one quantum. Stimulated emission involves the inverse process. The diagram also illustrates the requirement for conservation of energy that $\nu_{12} = \nu_m$ for an interaction to take place.

For spontaneous emission, (6.27) provides the surprising result that even when there is no field (n=0) there still remains a random 'zero point' energy of ½hν joules. Spontaneous emission is therefore a result of the interaction of the atom with this random 'zero point' field. There are further problems with the analysis, not least the fact that a radiation field may contain an infinite number of modes, and hence an infinite zero point energy (even when it does not exist!), but this problem is beyond the scope of the present book. For the moment, it can be circumvented by noting that the spontaneous emission is merely the result of the interaction between an atom and a single field mode.

6.9 STATES OF THE QUANTIZED RADIATION FIELD

The discussion of the previous section, although providing a formally satisfactory solution for the problem of the existence of spontaneous emission, leaves many questions unanswered, questions which only demanded an answer after the invention of the laser. Firstly, what is the relationship between the semi-classical analysis of the atom–field interaction outlined in sections 6.4–6.7 and the quantized field picture?

It turns out that this is no problem. In the quantum picture, the law of conservation of energy ensures that only interactions take place where the field gains or loses a quantum of energy exactly equal to that lost or gained by the atom (Fig. 6.4). In the semi-classical picture, from the infinite spectrum of all possible increases or decreases in field energy, the only one that is picked out is that which satisfies the law of conservation of energy. In fact, the mathematical formalisms for the semi-classical and fully quantized pictures are identical and all the results obtained in sections 6.5 and 6.6 are unchanged.

A second, more intractable problem concerns the physical nature of the electric and magnetic field vectors. In classical electromagnetism, it seems that these vectors have a real existence, invoked when lining up satellite dishes, TV aerials, etc. Apparently, in the quantum world such things cannot be done! This conceptual difficulty existed for thirty years after the original theoretical work on field quantization, but does not seem to have been seriously considered, possibly because quantum physicists and radio engineers did not socialize much. It surfaced dramatically with the invention of the laser. Laser radiation appears to a very good approximation to be a classical travelling wave of well-defined amplitude and phase. On the other hand, it is triumphantly the product of basic atomic physics. To compound the paradox, in the early days of laser theory the most comprehensive and wide-ranging theoretical model treated the radiation field classically (see Section 10.3).

The problem was resolved, not without difficulty, by returning to the basic quantum field theory. In 1965 Glauber pointed out that although a description of the quantized electromagnetic field in terms of photon number was very useful, it was not the only way it could be done.

The problem can be viewed from a different angle. For mechanical systems, such as the energy states of a simple harmonic oscillator, the quantum result 'grows over' to the classical result for macroscopic energies. Or to put it another way, for macroscopic energies, the energy spacing between quantized states is so small that the energy spectrum appears continuous. Since the description of the electromagnetic field in terms of the photon number operator does not have this property, it is necessary to define the field in terms of a different set of states. These states were first defined by Glauber as 'coherent states'. They are defined in terms of the photon number states and have the basic property that a measurement of the energy of a field in such a state does not yield a precisely defined number of quanta. Rather, successive measurements yield a Poisson probability distribution for the measured number of quanta, with an average number of quanta $\langle n \rangle$ and a spread in the distribution of $\langle n \rangle^{1/2}$. In other words, as the number of quanta in a mode becomes very large, the relative spread becomes very low. If the phase operator is considered, a similar situation occurs. In other words, coherent states are states where the product of the uncertainties of number and phase measurements decrease as the photon number increases, and the field becomes more and more like a classical wave with well-defined amplitude and phase. This is illustrated in Fig. 6.5.

The radiation produced by a continuous laser is a good approximation to a coherent state. On the other hand, light from 'spontaneous emission' sources such as gas discharge tubes involve a large number of independent emitters, so that

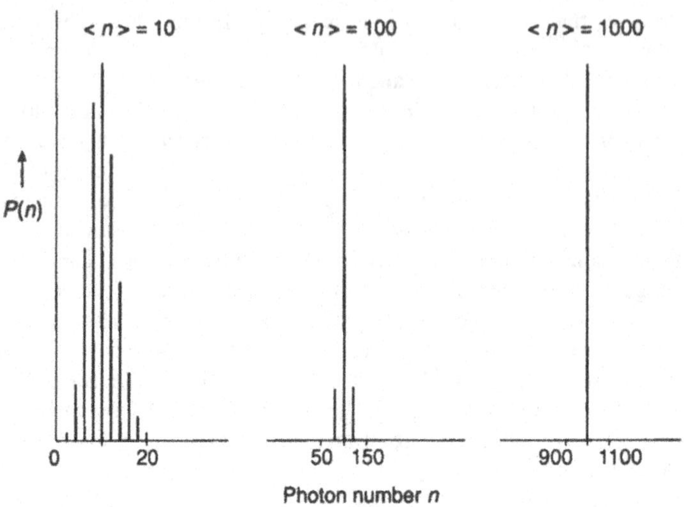

Fig. 6.5 Graphs of the relative probability of detecting n photons as a function of n for coherent states with mean photon numbers equal to 10, 100 and 1000. As the mean photon number increases, the relative spread decreases and the picture becomes more and more like that of a classical wave with well-defined amplitude and phase.

the statistical properties of the light are defined by the statistics of the emitters. In that case a large number of modes are excited, but the probability of a quantum occurring in any one mode is very small. In a laser, on the other hand, only one mode may be excited, so the average number of quanta in this mode is high.

Thus the properties of radiation from a gas laser are dramatically different from those generated by a conventional gas discharge tube, a subject which will be returned to in Chapter 10.

6.10 THE PHOTON

The previous sections have given a basic picture of the process of the interaction of atoms with electromagnetic radiation and have produced important results concerning allowed and forbidden transitions in atomic systems and the magnitude of transition probabilities. Sections 6.3 to 6.6 have developed the analysis in terms of the interaction between a classical field and a quantized atomic system. For conventional sources of radiation, the interaction energy is small compared with the energy an electron possesses by virtue of the attraction to the atomic nucleus, so perturbation theory produces quite acceptable results..

Sections 6.8 and 6.9 have extended the analysis by considering the quantization of the electromagnetic field. One important point arises here which must be emphasized. The idea of a 'photon' is only firmly based on the quantization process which results in (6.27). In elementary work, photons are often discussed as though they were small 'particles of light', well defined in space and indivisible. This simplistic picture is in no way valid, as can be seen from (6.27). When the photon number is zero there still appears to be a remaining 'half photon'.

This point is important when questions of the properties of radiation produced by conventional gas discharge tubes and gas lasers are discussed, particularly when experiments such as Young's double slit are in question. Photons can best be thought of in terms of quantized energy states of a mode of a radiation field. With this definition, it is plain that the physical dimensions of a photon are defined by the physical extent of the field mode itself.

One question remains: under what circumstances is it permissible to use semi-classical theory and when does the full quantum field theory have to be used? In general, at a first level it is acceptable to use semi-classical theory. As has been shown above, it is quite acceptable for the perturbational analysis of the interaction of atoms and fields. Similarly, when laser radiation is discussed, the radiation may, in the majority of cases, be considered as a classical beam of radiation with well-defined amplitude and phase (see Section 10.3) It must be stressed that in those cases semi-classical and quantum theory give identical results. However, in any cases when small photon numbers are encountered it is necessary to use the quantized field approach as effects occur which cannot be predicted by the classical analysis. This includes most obviously the spontaneous emission of radiation, but there are many other fascinating effects which are still the subject of research.

7

More complex atoms

Extending atomic theory to atoms more complex than hydrogen is simple in principle. Consider a model of the helium atom, two electrons in the field of a nucleus of charge $+2e$ (Fig. 7.1).

7.1 SCHRÖDINGER'S EQUATION FOR MORE COMPLEX SYSTEMS

The time-independent Schrödinger equation for the system can be written by extension from equation (3.14) as

$$\begin{aligned}[-(h^2/8\pi^2)(\nabla_1^2 + \nabla_2^2) - 2e^2/4\pi\varepsilon_0 | r_1 - 2e^2/4\pi\varepsilon_0 | r_2 \\ + e^2/4\pi\varepsilon_0\, r_{12} |]\, \Psi(1,2) = E\,\Psi(1,2).\end{aligned} \qquad (7.1)$$

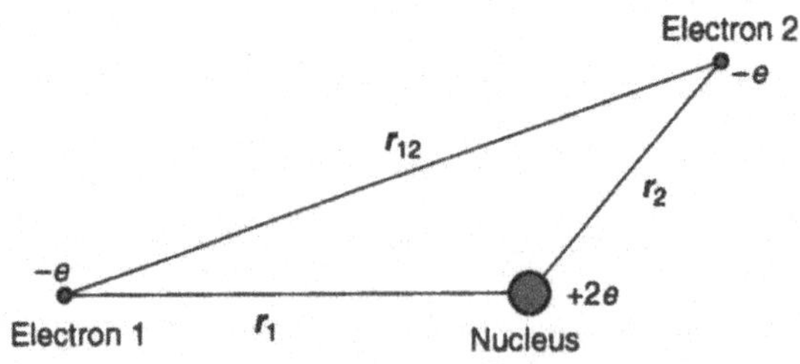

Fig.7.1 Model of the helium atom with two electrons in the field of a nucleus (sizes of electrons and nucleus much exaggerated).

This equation now includes the kinetic energies of the two separate electrons (labelled 1 and 2) and their potential energies due to the nucleus and to each other. The state function $\Psi(1,2)$ of the whole system is therefore a function of the co-ordinates of both electrons.

It is straightforward to extend (7.1) for atoms with three or more electrons. The problem, however, is to solve such an equation!

Nowadays these equations can be solved by computer without too much difficulty. However, the interpretation and picturing of the solutions is quite hard Because of this, it is still useful to use approximations in many cases, which simplify the problem considerably and enable the physical situations to be grasped more easily. For example, looking at (7.1), if the term describing the interaction between the two electrons (the fourth term in the square brackets) is disregarded, the equation can be separated into two 'hydrogen-like' equations, one for each electron.

Physically this approximation means assuming that the energy of each electron due to its interaction with the other is assumed to be much smaller than that due to its interaction with the nucleus. This may seem odd at first sight, but can be justified by the consideration that the charge on the nucleus is much greater than that on an individual electron and that, on average, the distance between different electrons is likely to be greater than that between an electron and the nucleus.

7.2 THE PAULI EXCLUSION PRINCIPLE

Given the separation of (7.1) into two independent equations, it is possible to think of 'hydrogen-like' eigenfunctions for each electron, $\psi_{n,l,m,s}(1)$ and $\psi_{n',l',m',s'}(2)$. (The spin quantum number s has been added here, although it does not feature in the solution of the spatial TISE.)

However, it must still be possible to write down the eigenfunction for the whole system, $\Psi(1,2)$, in terms of combinations of the single electron eigenfunctions. It turns out that there is only one way in which this can be done for electrons

$$\Psi(1,2) = [1/\sqrt{2}][\psi_{n,l,m,s}(1)\ \psi_{n',l',m',s'}(2)\ -\ \psi_{n',l',m',s'}(1)\ \psi_{n,l,m,s}(2)]. \qquad (7.2)$$

This is a special case of a general result, given in Appendix F. The state function is 'antisymmetric'; exchanging electrons produces a reversal of sign

$$\Psi(2,1) = -\ \Psi(1,2).$$

This has no immediate significance for the shape of the 'charge cloud' surrounding the nucleus, since the probability of finding electrons at any two specified points is given by $P(1,2) = |\Psi(1,2)|^2$, but there is one vital consequence. If the two electrons are in the same eigenstate ($n' = n$, $l' = l$, $m' = m$, $s' = s$), (7.2) shows that $\Psi(1,2)$ will be zero. Thus, such a state cannot exist.

This is a special case of the important principle known as the Pauli exclusion principle (see Appendix F). In an atom, no two electrons can be in the same eigenstate. In other words, in an atom, no two electrons can have the same set of quantum numbers (n,l,m,s).

7.3 THE AUFBAU PRINCIPLE AND ELECTRON CONFIGURATIONS

The Pauli exclusion principle enables the ground states of complex atoms to be modelled very simply by using the Aufbau (German for building-up) principle. This works as follows.

For an atom of atomic number Z, one starts with a 'naked' nucleus of charge Ze. Z electrons are then added, one at a time. Each electron will fall to its state of lowest energy. However, the Pauli exclusion principle means that states get progressively filled, from the lowest upward. The result is the 'electron configuration' of an atom in its ground state.

This configuration can be written down in terms of the principal quantum numbers and the spectroscopic symbols (s, p, d, etc) which denote the angular momentum quantum number. For example, the ground state configuration of hydrogen is (1s), that of helium is (1s1s) or $(1s)^2$ and that of lithium is $(1s)^2 2s$.

In this way, the electron configuration of every element can be built up. There is one complication. The approximation made in section 7.1 that the electron–electron interaction could be ignored, would mean that levels would fill up exactly in line with the hydrogen energy level diagram (Fig. 4.5). However, this approximation does not hold true for higher values of n. The measure of the goodness of the approximation is the spatial 'overlap' of the eigenfunctions. The greater the 'overlap', the greater is the effect of the electron–electron interaction. This interaction is repulsive, so it will tend to increase the energy of the state in proportion to the size of the overlap.

Thus, when states with $n = 3$ are considered, once the 3p states are filled, the next electron to be added might be expected to go into a 3d state. In fact, due to the overlap effect, the 3d state rests at a higher energy than the 4s, so the 4s states are filled first. The order in which the states are filled is

1s, 2s, 2p, 3s, 3p, 4s, 3d, 4p, 5s, 4d, 5p, 6s, 4f, 5d, 6p, 7s, 6d.

This sequence holds for the electron configurations of atoms with Z below about 20 and the electron configuration of the ground state of any element can be worked out. For larger atoms, more complex overlapping of states occurs. A table of electron configurations is given in Table 7.1. Note that we have not

Table 7.1 Table of elements from hydrogen (Z=1) to radium (Z=88) with their ground state electron configurations and ionization energies. Spaces mark 'filled shells'..

	Z	Configuration	E_I (eV)
H	1	1s	13.6
He	2	$(1s)^2$	24.6
Li	3	(He)2s	5.4
Be	4	$(He)(2s)^2$	9.3
B	5	$(He)(2s)^2 2p$	8.3
C	6	$(He)(2s)^2(2p)^2$	11.3
N	7	$(He)(2s)^2(2p)^3$	14.5
O	8	$(He)(2s)^2(2p)^4$	13.6
F	9	$(He)(2s)^2(2p)^5$	17.4
Ne	10	$(He)(2s)^2(2p)^6$	21.6
Na	11	(Ne)3s	5.1
Mg	12	$(Ne)(3s)^2$	7.6
Al	13	$(Ne)(3s)^2 3p$	6.0
Si	14	$(Ne)(3s)^2(3p)^2$	8.1
P	15	$(Ne)(3s)^2(3p)^3$	11.0
S	16	$(Ne)(3s)^2(3p)^4$	10.4
Cl	17	$(Ne)(3s)^2(3p)^5$	13.0
Ar	18	$(Ne)(3s)^2(3p)^6$	15.8
K	19	(Ar)4s	4.3
Ca	20	$(Ar)(4s)^2$	6.1
Sc	21	$(Ar)(4s)^2 3d$	6.5
Ti	22	$(Ar)(4s)^2(3d)^2$	6.8
V	23	$(Ar)(4s)^2(3d)^3$	6.7
Cr	24	$(Ar)(4s)^2(3d)^4$	6.7
Mn	25	$(Ar)(4s)^2(3d)^5$	7.4
Fe	26	$(Ar)(4s)^2(3d)^6$	7.9
Co	27	$(Ar)(4s)^2(3d)^7$	7.8
Ni	28	$(Ar)(4s)^2(3d)^8$	7.6
Cu	29	$(Ar)(4s)^2(3d)^9$	7.7
Zn	30	$(Ar)(4s)^2(3d)^{10}$	9.4
Ga	31	$(Ar)(4s)^2(3d)^{10}4p$	6.0
Ge	32	$(Ar)(4s)^2(3d)^{10}(4p)^2$	8.1
As	33	$(Ar)(4s)^2(3d)^{10}(4p)^3$	10.0
Se	34	$(Ar)(4s)^2(3d)^{10}(4p)^4$	9.8
Br	35	$(Ar)(4s)^2(3d)^{10}(4p)^5$	11
Kr	36	$(Ar)(4s)^2(3d)^{10}(4p)^6$	14.0

	Z	Configuration	E_I (cV)
Rb	37	(Kr)5s	4.2
Sr	38	$(Kr)(5s)^2$	5.7
Y	39	$(Kr)(5s)^2 4d$	6.6
Zr	40	$(Kr)(5s)^2(4d)^2$	7.0
Nb	41	$(Kr)(5s)^2(4d)^3$	6.8
Mo	42	$(Kr)(5s)^2(4d)^4$	7.2
Tc	43	$(Kr)(5s)^2(4d)^5$	?
Ru	44	$(Kr)(5s)^2(4d)^6$	7.5
Rh	45	$(Kr)(5s)^2(4d)^7$	7.7
Pd	46	$(Kr)(5s)^2(4d)^8$	8.3
Ag	47	$(Kr)(5s)^2(4d)^9$	7.6
Cd	48	$(Kr)(5s)^2(4d)^{10}$	9.0
In	49	$(Kr)(5s)^2(4d)^{10}5p$	5.8
Sn	50	$(Kr)(5s)^2(4d)^{10}(5p)^2$	7.3
Sb	51	$(Kr)(5s)^2(4d)^{10}(5p)^3$	8.6
Te	52	$(Kr)(5s)^2(4d)^{10}(5p)^4$	9.0
I	53	$(Kr)(5s)^2(4d)^{10}(5p)^5$	10.4
Xe	54	$(Kr)(5s)^2(4d)^{10}(5p)^6$	12.1
Cs	55	(Xe)6s	3.9
Ba	56	$(Xe)(6s)^2$	5.2
	57–71	(Rare Earths)	
Hf	72	$(Xe)(6s)^2(4f)^{14}(5d)^2$	5.5
Ta	73	$(Xe)(6s)^2(4f)^{14}(5d)^3$	7.9
W	74	$(Xe)(6s)^2(4f)^{14}(5d)^4$	8.0
Re	75	$(Xe)(6s)^2(4f)^{14}(5d)^5$	7.9
Os	76	$(Xe)(6s)^2(4f)^{14}(5d)^6$	8.7
Ir	77	$(Xe)(6s)^2(4f)^{14}(5d)^7$	9.2
Pt	78	$(Xe)(6s)^2(4f)^{14}(5d)^8$	9.0
Au	79	$(Xe)(6s)^2(4f)^{14}(5d)^9$	9.2
Hg	80	$(Xe)(6s)^2(4f)^{14}(5d)^{10}$	10.4
Tl	81	(Hg)6p	6.1
Pb	82	$(Hg)(6p)^2$	7.4
Bi	83	$(Hg)(6p)^3$	7.3
Po	84	$(Hg)(6p)^4$	8.4
At	85	$(Hg)(6p)^5$	?
Rn	86	$(Hg)(6p)^6$	10.7
Fr	87	(Rn)7s	?
Ra	88	$(Rn)(7s)^2$	5.3

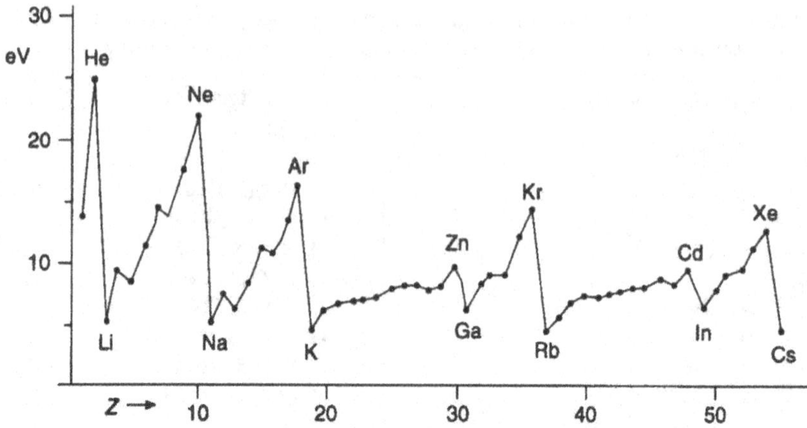

Fig. 7.2 Ionization energies of the elements plotted as a function of atomic number Z up to Z=55. Maxima are very non-reactive elements, minima are very reactive elements.

been able to say anything about excited states of the atoms. One important experimental quantity is relatively easily measured – the ionization energy. This is the minimum energy required to remove one electron from the atom, and will therefore be the energy needed to remove the least tightly bound electron, the electron with the highest value of n. Measured ionization energies are included in Table 7.1 and a graph of ionization energy versus Z is given in Fig. 7.2.

7.4 THE PERIODIC TABLE

In 1869–71 Mendeleev arranged the known elements in a 'periodic table' in terms of their chemical properties. A modern version of this table is shown in Table 7.2.

It is surprisingly easy to use the electron configurations of Table 7.1 and the ionization energy graph of Fig. 7.2 to match the chemical properties of the various groups in the table. For example, Group VIII contains the chemically inert 'noble gases'. Table 7.1 shows the electron configurations of these elements to be 'closed shells'. This means that for each value of n, all possible states of l, m and s contain electrons. For the spatial part of the state function, this gives the picture of a spherically symmetric cloud of charge (recall section 3.4). It is a reasonable step to assume that such a charge distribution around the nucleus is very stable – hence the noble gases have high ionization energies and are not chemically reactive. As might be expected, the ionization energy decreases with increasing Z. The noble gases correspond to the peaks in Fig. 7.2.

It is worth noting for future reference that the spherical symmetry of such a closed shell means that the total orbital angular momentum of the electrons is zero, as is the total spin angular momentum.

Following the train of argument, the next interesting group is Group I. In this case a single electron is added to a closed shell, causing a dramatic change in ionization energy and chemical properties. The single electron 'sees' a nucleus surrounded by a spherical shell of negative charge. It is therefore far less tightly bound to the nucleus. Its ionization energy is low and it is highly chemically reactive. Compare potassium (Z = 19) with its neighbouring element argon (Z=18). Furthermore, the spectra of such elements have line series like those of hydrogen.

Elements in Group III also have spherically symmetric arrangements for their inner electrons and a single outer electron. However, in these cases the outer electron is in a p state. Nevertheless, they have some properties similar to those of Group I elements.

Interestingly, the elements in Group VII have similar properties to those in Group III. These elements have filled inner shells and outer shells filled, with the exception of a single electron. They have properties which suggest that they act like elements with filled shells and a single positive 'hole' playing the role of the floating electron in Group III elements.

Of the remaining groups, Group II elements have closed shells with two s electrons outside them. By analogy with Group I, these elements behave spectroscopically like helium, with the central positive charge shielded.

Group VI elements bear the same relationship to Group II as Group VII elements do to Group III. Rather than the single floating 'hole' of the Group VII element, there are two 'holes' which, apart from sign, act like two electrons interacting with a filled shell.

Group IV – containing the important element carbon – has a spherically symmetric charge cloud with two p electrons outside it. The consequences of this configuration in terms of the spectroscopy of carbon will be discussed later in Section 9.3.

Finally, Group V elements have three floating electrons outside closed subshells. They are complicated!

The Aufbau principle has enabled a considerable amount of sense to be made of the originally empirical periodic table. There are some cases that have been ignored. For example, elements with atomic numbers 26, 27 and 28 (iron, cobalt and nickel) are assigned to Group VIII, even though their configurations are not, at first sight, very similar to those of the noble gases. These cases, together with the quantitative discussion of energy levels and transitions obviously require much more detailed calculation and the removal of the assumption that electron–electron forces may be considered negligible. However, as will be seen, it is possible to go a surprisingly long way, using relatively simple concepts, in reaching a qualitative understanding the properties of the chemical elements.

Table 7.2 Periodic Table of the elements. Atomic weights (to 4 significant figures) are based on taking carbon as 12. The elements are grouped in columns. Rare earth and actinide elements are not included.

I	II	III	IV	V	VI	VII	VIII
1 H 1.008							2 He 4.003
3 Li 6.940	4 Be 9.012	5 B 10.81	6 C 12.00	7 N 14.01	8 O 16.00	9 F 19.00	10 Ne 20.17
11 Na 22.99	12 Mg 24.31	13 Al 26.98	14 Si 28.09	15 P 30.97	16 S 32.06	17 Cl 35.45	18 Ar 39.95
19 K 39.09	20 Ca 40.08	21 Sc 44.96	22 Ti 47.90	23 V 50.94	24 Cr 52.00	25 Mn 54.94	26 Fe 27 Co 28 Ni 55.85 58.93 58.71
29 Cu 63.55	30 Zn 65.38	31 Ga 69.72	32 Ge 72.59	33 As 74.92	34 Se 78.96	35 Br 79.90	36 Kr 83.80
37 Rb 85.47	38 Sr 87.62	39 Y 88.91	40 Zr 91.22	41 Nb 92.90	42 Mo 93.94	43 Tc 98.91	44 Ru 45 Rh 46 Pd 101.1 102.9 106.4
47 Ag 107.9	48 Cd 112.4	49 In 114.8	50 Sn 118.7	51 Sb 121.7	52 Te 127.6	53 I 126.9	54 Xe 131.3
55 Cs 132.9	56 Ba 137.3	57-71 Rare Earths	72 Hf 178.5	73 Ta 180.9	74 W 183.9	75 Re 186.2	76 Os 77 Ir 78 Pt 190.2 192.2 195.1
79 Au 197.0	80 Hg 200.6	81 Tl 204.4	82 Pb 207.2	83 Bi 209.0	84 Po (209)	85 At (210)	86 Rn (222)
87 Fr (223)	88 Ra 226.0	89-103 Actinides					

8

The helium atom and term symbols for complex atoms

Atomic electron configurations give basic information about the stable atom in its ground state. For more detail, it is necessary to return to experimental spectral analysis. This can most easily be approached via the analysis of helium.

8.1 SPECTRA AND ENERGY LEVEL DIAGRAMS FOR HELIUM

Helium, the next element after hydrogen, has a line spectrum similar to that of hydrogen (Fig. 8.1). A term analysis of the spectral lines produces an energy level diagram which resembles that of hydrogen (Fig. 8.2). As with hydrogen, each level is categorized by a principal quantum number n and an angular momentum quantum number l, where l may have values $n - 1, n - 2, \ldots, 0$.

This ties in with the electron configuration analysis of the last chapter. If each electron has hydrogen-like eigenstates, it can be assumed that each state is defined by a set of values of n, l and m. The ground state will have both electrons with $n = 1$ and $l = m = 0$. The first excited state will have one electron with $n = 1$ and the other with $n = 2$, the next state has one with $n = 1$ and the

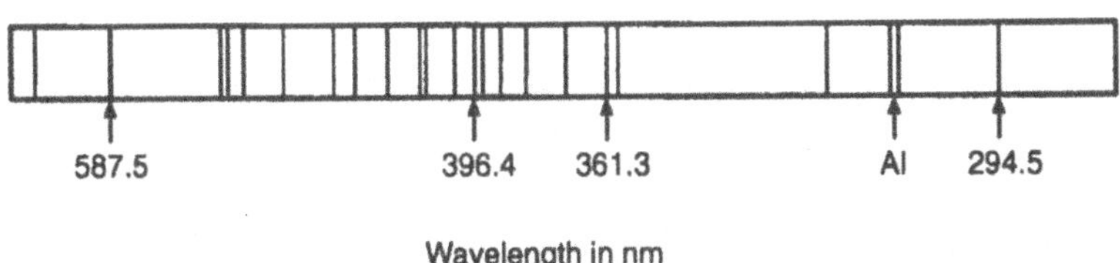

Fig. 8.1 Helium spectrum using a quartz spectrometer. Two lines from the spectrum of aluminium are also present, presumably from the discharge tube electrodes.

other with $n = 3$, etc. It will be seen later (8.3) that these pairings of values of n give the lowest lying energy states, so the values of n on the energy level diagram refer to one electron, while it is assumed that the other electron remains in the $n = 1$ state.

However, there is one surprising result. The energy levels divide themselves into two distinct groups and no radiation is observed corresponding to transitions between the groups. This led early researchers to conclude that there must be two types of helium, which they named para-helium and ortho-helium. The emission lines in the two groups follow the selection rules worked out previously, but the lowest lying state of ortho-helium, which corresponds to a value of $n = 2$, lies well above the ground state ($n = 1$), which only occurs in

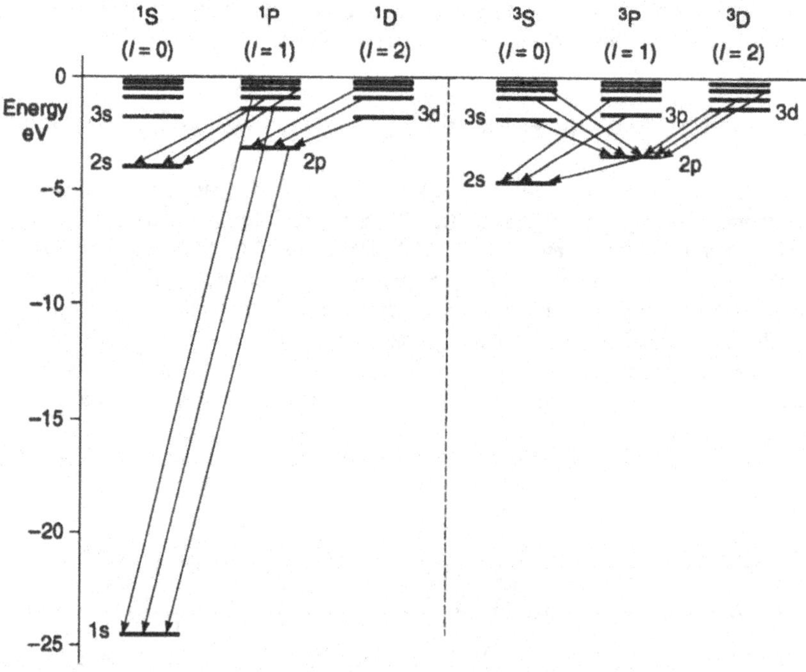

Fig. 8.2 Energy level diagram (sometimes called a term diagram) for helium. The singlet and triplet states are shown separately and the appropriate principal quantum number is given in some cases. Since one electron always has $l = 0$, the term symbols (shown at the top) are particularly simple in this case. Some of the allowed transitions are also indicated. The $2s^1S$ and $2s^3S$ states are both metastable.

para-helium. This lowest lying state of ortho-helium is therefore referred to as a metastable state.

Two problems therefore arise: first, the reason for the strange grouping of energy levels (since there are obviously not two types of helium) and second, the calculation of the absolute values of the energy levels.

8.2 EIGENFUNCTIONS FOR THE HELIUM ELECTRONS – SINGLET AND TRIPLET STATES

The answer to the first problem comes from the analysis which led to the definition of the state function $\Psi(1,2)$ in Chapter 7, and hence to the Pauli exclusion principle. Each electron has a spatial eigenfunction ($u(1)$ for electron 1, $u(2)$ for electron 2) and, independently, a spin eigenfunction ('spin up', α, or 'spin down', β). The function $\Psi(1,2)$ must be made up of combinations of these eigenfunctions.

Following the analyses in Appendices F and G, there is a further restriction that $\Psi(1,2)$ must be antisymmetric

$$\Psi(1,2) = -\Psi(2,1)$$

and therefore $\Psi(1,2)$ can be produced in four ways.

First, making the spatial part symmetric and the spin part antisymmetric gives one possible combination

$$\Psi(1,2) = [u(1)\,u'(2) + u'(1)\,u(2)\,]\,[\alpha(1)\beta(2) - \beta(1)\alpha(2)]\,/2. \qquad (8.1)$$

Second, making the spatial part antisymmetric and the spin part symmetric gives three possible combinations:

$$\begin{aligned}\Psi(1,2) &= [u(1)\,u'(2) - u'(1)\,u(2)\,]\,\alpha(1)\alpha(2)/(\sqrt{2})\\ \Psi(1,2) &= [u(1)\,u'(2) - u'(1)\,u(2)\,]\,\beta(1)\beta(2)\,/(\sqrt{2})\\ \Psi(1,2) &= [u(1)\,u'(2) - u'(1)\,u(2)\,]\,[\alpha(1)\beta(2) + \beta(1)\alpha(2)]\,/2.\end{aligned} \qquad (8.2)$$

So for a situation where electron 1 is in a state defined by the quantum numbers (n, l, m and s) and electron 2 is in a state defined by the quantum numbers (n', l', m' and s') there are four possible eigenstates. The state defined by (8.1) is described as a 'singlet' state and those by (8.2) as 'triplet' states, for obvious reasons

Two important results follow immediately. First, the total spin of the electrons in a singlet state is zero while that in a triplet state is 1. A transition between a singlet and triplet states would thus involve a change in the total spin of the system. This cannot be produced by the normal interaction with an electromagnetic field, so such transitions are forbidden.

Second, the ground state of a helium atom involves two electrons, each in a $n = 1$ state. From (8.2), therefore, there can be no triplet states since in that case the spatial part of the statefunction is zero. Thus the ground state of helium must be a singlet state and the lowest possible triplet state (where $n = 1$ and $n' = 2$) will not be connected to the ground state via an optical transition. It will therefore be a 'metastable' state. An atom in such a state can lose energy and return to the ground state (for instance by collision with the walls of the container or with another atom). However, the probability of this happening is low, so that atoms may remain in metastable states for milliseconds or longer, 10^5 times longer than the time spent in an excited state that is optically connected to the ground state (note that the excited singlet state where $n' = 2$ and $l = 0$ will also be metastable, due to the normal selection rules).

Thus the division into triplet and singlet states is explained as another consequence of the Pauli exclusion principle.

8.3 ENERGY EIGENVALUES

Using the assumption of hydrogen-like eigenstates, the energy eigenvalues can be simply written down using (3.16)

$$E_{nn'} = [-me^4/2\varepsilon_0{}^2h^2]\,[1/n^2 + 1/n'^2] \quad \mathrm{J}. \tag{8.3}$$

Examination of this equation shows immediately that states where $n = 1$ and n' has any integral value lie lower in energy than states with any other combination of n and n'.

A comparison of the theoretical values for $E_{nn'}$ given by (8.3) with the experimental results produces only fair agreement (Fig. 8.3). In general the energies predicted by (8.3) lie much lower than those found experimentally. This, of course, is due to neglect of the electron–electron repulsion which would raise the energy of each eigenstate.

Of course, the simple theory gives eigenenergies independent of the angular momentum quantum number and of the multiplicity of the state. Experimentally, it can be seen that this energy degeneracy breaks down and that the breakdown occurs in a well-defined way.

First, when either singlet or triplet states are considered, states with a configuration 1s2s lie lower in energy than those with configuration 1s2p. Similarly 1s3s lies lower than 1s3p which lies lower than 1s3d, etc. Second, for a particular configuration, triplet states lie lower than singlet states.

This is due to the electron–electron interaction. For a better fit between theory and experiment, this interaction has to be included in the calculation.

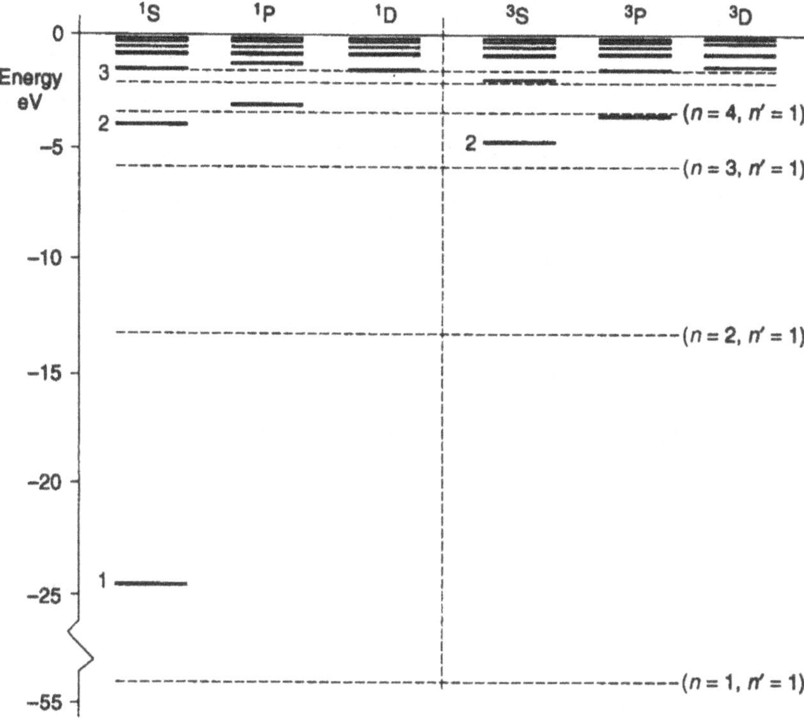

Fig. 8.3 Comparison of experimental helium energy levels (solid lines) with those predicted by simple theory, neglecting electron–electron interactions (dashed lines). It can be seen that the quantitative agreement is not good for small values of n.

8.4 THE EFFECT OF THE ELECTRON–ELECTRON INTERACTION

The obvious way to include this interaction is to return to the full TISE for the helium atom, (7.1). However, the comparison of theory with experiment in Fig. 8.3 suggests an easier approach. The effect of the electron–electron interaction is not too large, so the method of time-independent perturbation theory may again be used.

The correction to an energy eigenvalue due to a small energy perturbation was quoted in (5.11). For an extra energy term W, the correction to the energy eigenvalue E_n with equivalent eigenfunction u_n, is given by

$$\Delta E = \int u_n{}^* W \, u_n \, dv \quad \text{J}. \tag{8.4}$$

The integration is over all space and in this case, $W = e^2/4\pi\varepsilon_0 |r_{12}|$.

The eigenfunctions are given by the spatial part $U(1, 2)$ of $\Psi(1,2)$, as written in (8.1) and (8.2). To simplify: for a state where the two principal quantum numbers are p and q

$$U(1,2) = u_p(1)u_q(2) \pm u_q(1)u_p(2). \tag{8.5}$$

The positive sign is appropriate for a singlet state and the negative sign for a triplet state.

Substituting into (8.4) and multiplying out gives four terms

$$\begin{aligned} \Delta E &= \int u_p^*(1)\, u_q^*(2) W u_p(1)\, u_q(2) \mathrm{dv} + \int u_q^*(1)\, u_p^*(2) W u_q(1)\, u_p(2) \mathrm{dv} \\ &\pm \left[\int u_q^*(1)\, u_p^*(2) W u_p(1)\, u_q(2) \mathrm{dv} + \int u_p^*(1)\, u_q^*(2) W u_q(1)\, u_p(2)\, \mathrm{dv} \right]. \end{aligned} \tag{8.6}$$

The first two terms are common to both singlet and triplet states. Inspection shows that they actually correspond to the extra energy due to electrostatic repulsion between the electrons. Because of this, the extra energy is usually known as the Coulomb energy and will be greater the greater the overlap between the state functions $u_p(1)$ and $u_q(2)$. Consideration of the shape of hydrogen eigenfunctions (Fig. 3.3) shows this overlap to be greater for a pair of electrons in 1s and 2p states than for electrons in 1s and 2s states. This explains the first experimental result mentioned in 8.3.

The second two integrals in equation (8.6) have no classical electrostatic analogue, and the energy correction due to them is known as the exchange energy.

These terms are positive for singlet states and negative for triplet states. Hence, in general, for the same electron configuration, triplet states lie lower in energy than singlet states.

The electron–electron interaction causes no energy splitting between the three triplet states. However, as in the case of hydrogen, spin–orbit coupling adds an extra small energy term so that the spectral lines due to triplet–triplet transitions exhibit fine structure.

8.5 TERM SYMBOLS AND FINE STRUCTURE

It has been shown that a particular electron configuration can give rise to a number of possible states, which may have different energies and may have different allowed transitions. It is important to be able to label these states unambiguously and the label used is known as the 'term symbol' for the state in question. For many (but not all) of the elements, the term symbol incorporates all the information required.

Consideration of the possible states in helium suggests that the important parameters defining the state are the orbital angular momentum (OAM), the total spin angular momentum and the multiplicity. The term symbol incorporates information on all these.

First the total orbital angular momentum is defined. This is given by the quantum number L, defining the sum of the angular momenta l for the individual electrons. In helium this is particularly simple as L merely equals the value of l for one electron, since the second one always has $l = 0$. A letter is then used to define the value of L. These are identical with the letters used for the single electron in hydrogen, except that since the total OAM is specified, the letters are capitalized (S represents $L = 0$; P represents $L = 1$ etc). Second, the multiplicities of the state are specified by a superscript preceding the letter. For example, this terminology enables the first few states of helium to be written as follows (for completeness, the electron configurations are included)

Ground state: (1s1s) ^{1}S (pronounced 'one-s, one-s; singlet-S')
Excited states: (1s2s) ^{1}S or (1s2s) ^{3}S (pronounced 'triplet-S')
(1s2p) ^{1}P or (1s2p) ^{3}P (pronounced 'triplet-P')
etc.

Finally, the state of the total spin angular momentum, S must be specified. As this only takes real importance in the calculation of spin–orbit coupling energies, and as the calculation of this energy depends on the total angular momentum J (see Section 5.3), it is J that is specified (it should be noted that the multiplicity of the state is equal to $(2S + 1)$). The value of J is put as a subscript to the right of the letter. Adding this to the definitions above gives a complete definition of possible states.

Ground state: (1s1s) $^1S_0$ ('one-s, one-s; singlet-S-zero')
Excited states: (1s2s) $^1S_0$
(1s2s) $^3S_1$ ('singlet-S-one')
(1s2p) $^1P_1$
(1s2p) $^3P_0$
(1s2p) $^3P_1$
(1s2p) $^3P_2$
etc.

Using the previous discussion of spin-orbit coupling, it can be seen that the ^{3}P (pronounced 'triplet-P') terms will be split, with $^3P_0$ having the lowest energy and $^3P_2$ the highest.

8.6 HUND'S RULES FOR ENERGY LEVEL SPLITTING

The results for the relationships between energies of the different states in helium are special cases of general rules known as Hund's rules.

Hund's first rule states that for a given electron configuration, the term with maximum multiplicity falls lowest in energy. This merely follows from the appearance of the exchange energy in the electron–electron interaction energy.

Hund's second rule states that for a given multiplicity, the term with the largest value of L lies lowest in energy (e.g. $^3D < {}^3P$). This follows from the fact that the Coulomb energy is greater the greater the overlap between state functions.

Hund's third rule states that for a given state and multiplicity, the state with lowest value of J has lowest energy. This is due to spin–orbit energy as noted above. However, for elements in Group VI and Group VII, where the electron configuration appears to act like one or two positive 'holes' in a filled shell, the sign of the effective spin orbit energy is inverted and so the state of **highest** J has the lowest energy.

These general rules are mainly useful as a reminder of the processes involved. Examples of their application will appear in Chapter 9.

9

Further discussion of atomic structure

It is now possible to discuss in more detail some of the groups in the periodic table, using the term symbol formulation described in Section 8.5.

9.1 HYDROGEN-LIKE ATOMS

Group I of the table consists of hydrogen and other atoms whose configurations

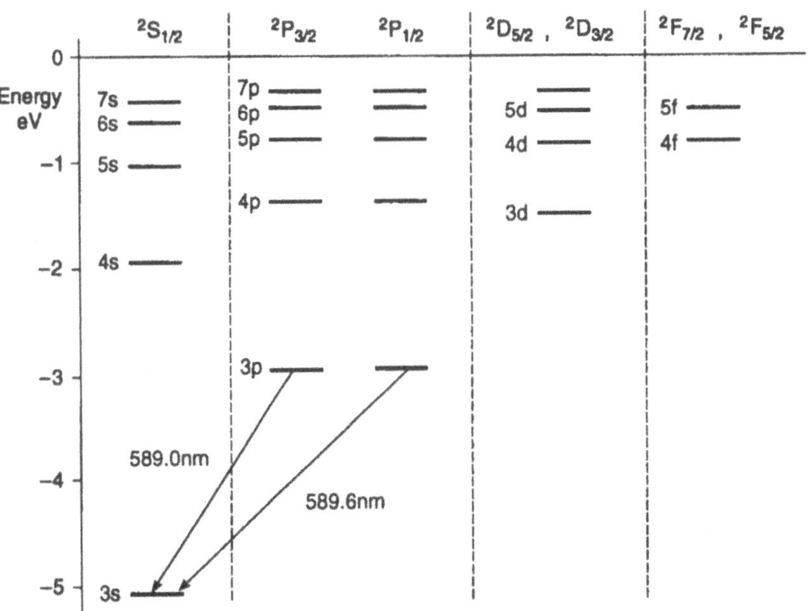

Fig. 9.1 Energy level diagram of sodium (group I). Term symbols are given at the top and the configuration of the 'active' electron is shown. The transitions producing the sodium yellow doublet are also marked.

consist of a closed, spherically symmetrical shell with a single electron outside. Important elements are Li (Z=3), Na (Z= 11) and K (Z= 19).

For all these atoms the ground state term symbol is $^2S_{1/2}$. The spectra and energy level diagrams are like those of hydrogen, allowing for the effect of the interaction of the single electron with the inner electron 'charge cloud' (Hund's second rule). An example is the spectrum and energy level diagram of sodium (Fig. 9.1).

9.2 HELIUM-LIKE ATOMS

Group II of the periodic table contains atoms which have a helium-like structure, with two s electrons outside a closed shell. Obvious examples are Be (Z = 4) and

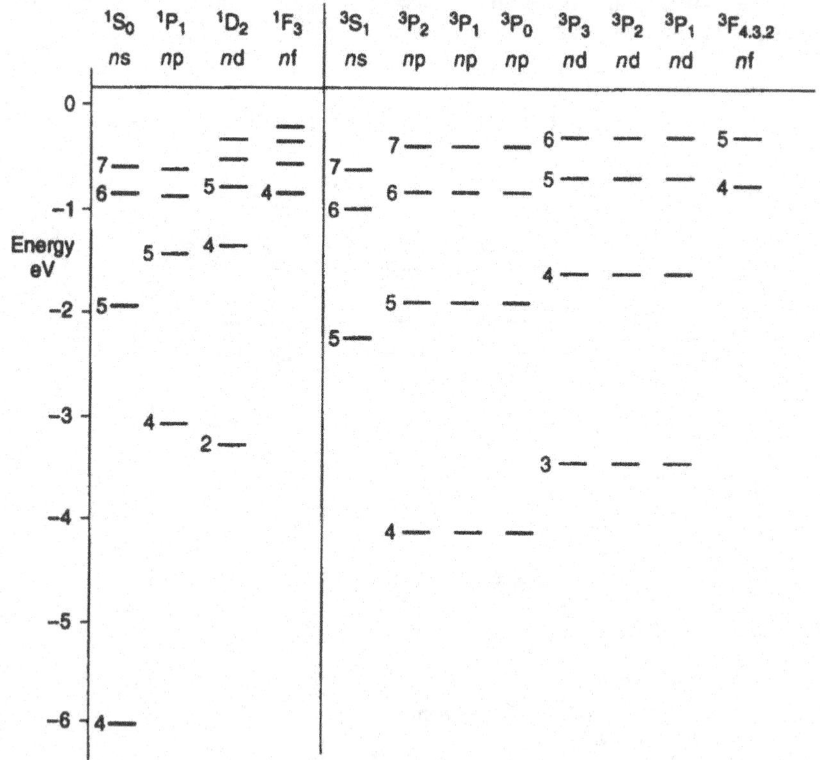

Fig. 9.2 Energy level diagram of calcium (group II). The singlet and triplet terms are separated out. At this resolution the energy difference between (for example) the ^3P terms cannot be seen.

Ca ($Z = 20$). In each case the ground state term symbol is $^1S_0$. The energy level diagram shows a division into singlet and triplet terms like helium(Fig. 9.2).

9.3 THE STRUCTURE OF CARBON

Carbon (group IV) is a very important element, so its structure will be discussed separately. Its electron configuration is $(1s)^2(2s)^2(2p)^2$ so the effective electrons will be the two (2p) electrons outside a spherical charge cloud. The energy level diagram is complex (Fig. 9.3), but it is possible to work out the ground state term.

This involves three steps. Firstly, it can be seen by vector addition that the

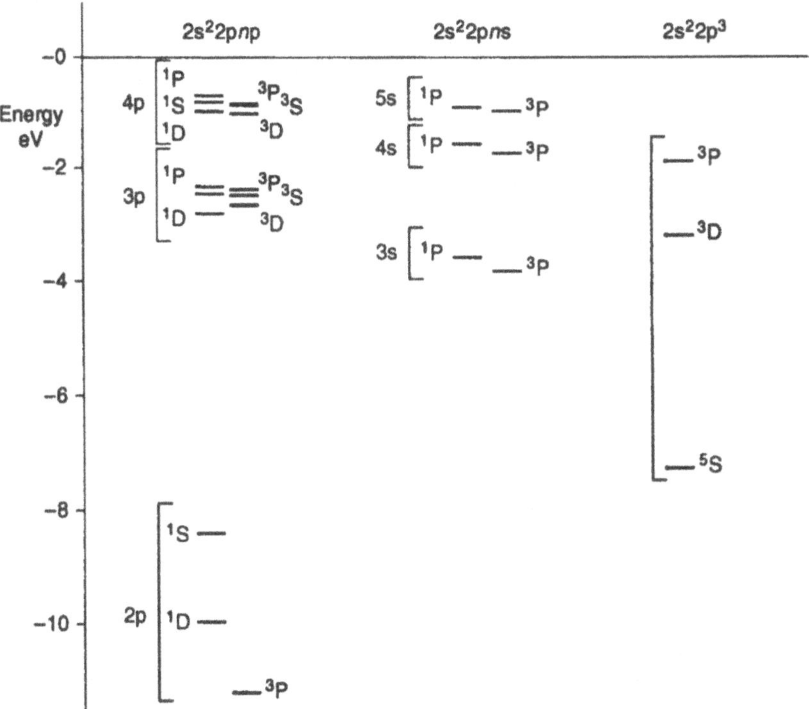

Fig. 9.3 Simplified energy level diagram of carbon (group IV). In this case the levels are sorted out into separate columns defined by electron configuration. The terms are differentiated within each column.

total orbital angular momentum L for this configuration can take values 0, 1 or 2. Similarly the total spin angular momentum S, can take the values 0 or 1.

Secondly, consider the states where $S = 0$. Since this means that the spin part of the eigenfunction is antisymmetric, the space part must be symmetric and this implies that L must take the values 0 or 2, leading to two possible states: $^1S_0$ or $^1D_2$. Conversely, states where $S = 1$ must must have $L = 1$ and this leads to states $^3P_2$, $^3P_1$ or $^3P_0$.

The final question is which of these five possible states has the lowest energy. Hund's first rule cuts out $^1S_0$, Hund's second rule cuts out $^1D_2$ and Hund's third rule gives $^3P_0$ as the state of lowest energy.

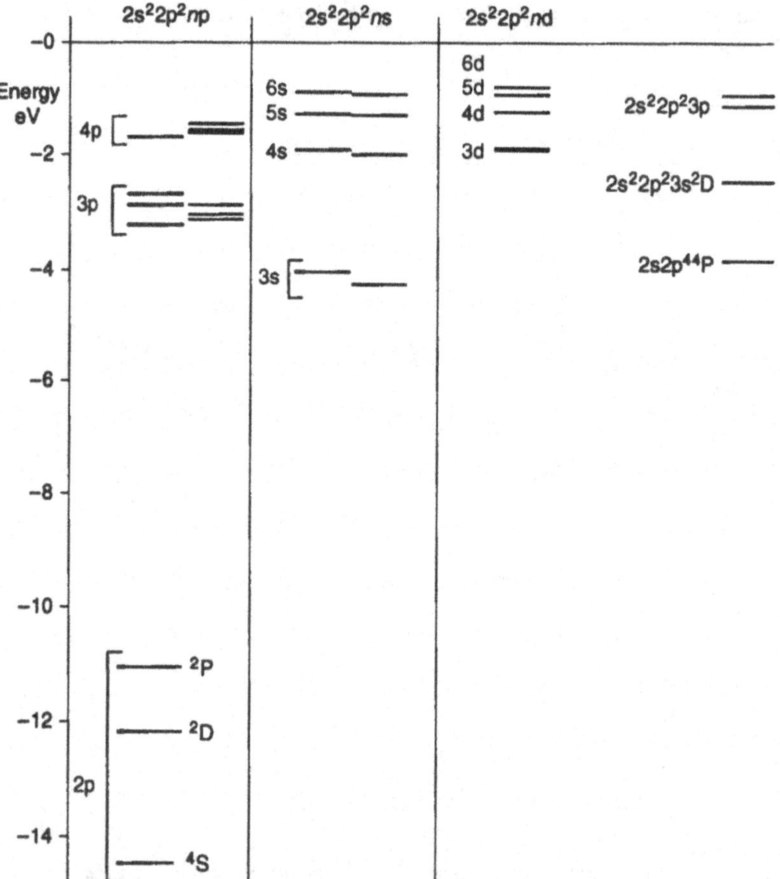

Fig. 9.4 Energy level diagram of nitrogen. In this case the configuration $1s^22s^22p^3$ produces three terms, of which the ^4S term is the lowest in energy.

9.4 OTHER ELEMENTS

Other elements with relatively low values of Z follow the pattern shown above, although detailed calculation of the term for the ground state may be more complex. For example, nitrogen with $Z = 7$ has its three most energetic electrons in the 3p state. Hund's third rule gives the spin of the ground state to be $^3/_2$. The spins of all three electrons must be parallel, so the spin part of the wave function must be symmetrical. Detailed consideration of possible orbital states shows that the only possible totally antisymmetric state is with $L = 0$, so the ground state is $^4S_{3/2}$ (Fig. 9.4).

For oxygen (Z=8), the outermost electrons are $(3p)^4$ and Hund's rule for shells that are over half full comes into play. The 'two-hole' configurations will be like

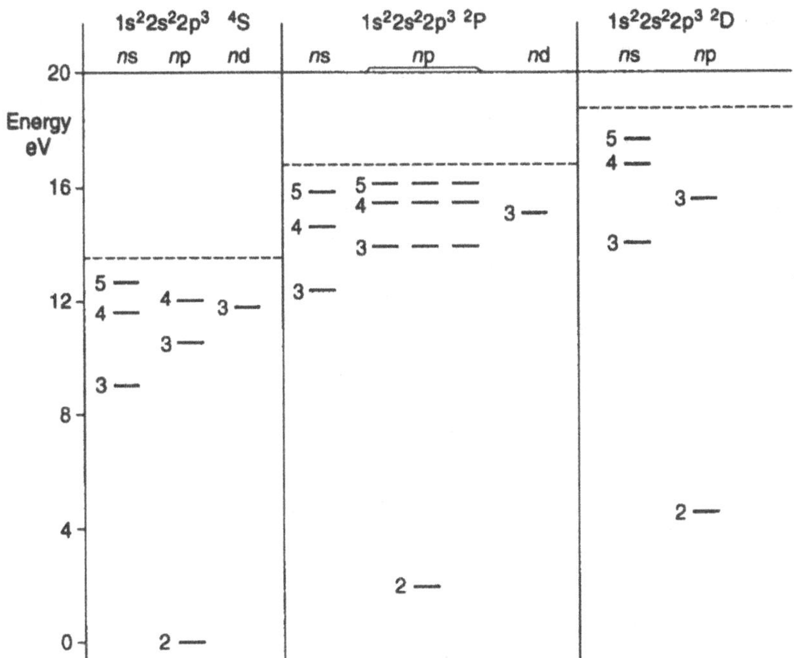

Fig. 9.5 Simplified energy level diagram for oxygen. In this case the Aufbau principle can be used to add a single electron to one of the three electron configurations equivalent to the lowest states in nitrogen (Fig. 9.4). Three series are produced, each with a different series limit (the energy scale is plotted upwards from the ground state in this case).

those of carbon with the difference that since the signs of the energy terms are reversed, so the ground state term is $^3P_2$ (Fig 9.5).

For higher values of Z, similar considerations can be used to work out the ground state terms. However there are very many detailed points which have to be taken into consideration, making calculation very difficult in most cases.

9.5 THE CALCULATION OF ATOMIC ORBITALS

So far the eigenfunctions for electrons in more complex atoms have been thought of merely in terms of the simple hydrogen orbitals of (3.17) and Table 3.1. These give a useful qualitative picture so long as it is realized that the actual eigenfunctions may diverge considerably from the values given for hydrogen, since each atomic electron finds itself moving in a field of force not due solely to the atomic nucleus, but also to the other atomic electrons. It is possible to calculate the actual form of the eigenfunction in any particular case but, as will be seen shortly, this procedure is lengthy. It is therefore useful to have an expression for the eigenfunction which gives a reasonable approximation for its shape. This can be achieved by using a version of the hydrogen-like eigenfunction with adjustable parameters specified by empirical rules depending on the situation of the particular electron considered. This procedure is rather like the use of Hund's rules in Section 8.6 and results in a set of eigenfunctions which give a reasonable approximation to the exact eigenfunction. The most useful such formulation gives what are known as 'Slater orbitals'.

The expression for a Slater orbital for an electron in an atom of atomic number Z is given by the following equation which bears considerable similarity to the hydrogen eigenfunction

$$u_{n,l,m}(r,\theta,\phi) = Cr^{\,n^*-1}\exp(-Z^*r/n^*r_o)\,Y_{l,m}(\theta,\phi). \tag{9.1}$$

In this case n, l and m are the usual quantum numbers and r_o is the Bohr radius. C is a normalizing constant and $Y_{l,m}$ is the appropriate spherical harmonic as defined in Table B.1. However, the expression also includes two parameters n^* and Z^*.

The **effective principal quantum number** n^*, is related to the principle quantum number n as shown below

$n = 1$ to 3 $\quad n^* = n$
$n = 4 \quad n^* = 3.7$
$n = 5 \quad n^* = 4$
$n = 6 \quad n^* = 4.2.$

The **effective atomic number** Z^* can be understood physically by the fact that a particular electron is attracted to the nucleus, but is also repelled by electrons closer than itself to the nucleus. Assuming that on average the force due to these other electrons can be thought of in terms of that due to a spherical cloud of negative charge surrounding the nucleus, the electron is shielded from the nucleus and the effective atomic number Z^* is less than Z. Normally

$$Z^* = Z - \sigma$$

where σ is called the shielding constant.

In order to calculate σ for a particular case, the shapes of the orbitals involved in the shielding have to be taken into account. This is done by the following procedure.

Orbitals are divided into the following groups

1s; 2s,2p; 3s,3p,3d; 4s,4p; 4d; 4f; 5s,5p; 5d.

This separation into groups can be compared with the Aufbau level filling of Section 7.3. It can be seen that in this case it is not the eigenenergy that is under consideration, but the shape of the eigenfunction, defined by the θ and ϕ components of the eigenfunction.

The value of σ may be calculated for an electron in a particular orbital by the following rules which reflect roughly the relative importance of each other electron in the shielding of the electron of interest.

First, effects due to electrons in higher groups may be ignored, since the amount of time they spend closer to the nucleus than the electron of interest is relatively small.

If the electron of interest is in the 1s state, $\sigma = 0.3$.

However, if the electron is in a higher state, σ is calculated by summing effects due to other electrons in the same group; effects due to electrons in the

Table 9.1 Examples of calculated values of Z^* for electrons in the ground states of different atoms

	H	Li	B	Na	Al
1s	1.0	2.7	4.7	10.7	12.7
2s		1.3	2.6	6.85	8.85
2p			2.6	6.85	8.85
3s				2.20	3.5
3p					3.5

next inner group, and finally electrons in all other inner groups.

For each other electron in the same group a value of 0.35 is added. For the group inside the one of interest, 0.85 is added if the electron of interest is in an s or p state, otherwise 1 is added.

For each electron in inner groups, 1 is added to σ.

Thus, for example, Z^* for a ground state electron in helium is 1.7, and for the outermost (3s) electron in Na, $Z^* = 2.2$. Table 9.1 gives values of Z^* for the various electrons in some atoms.

Having established these values, it is now possible to return to (9.1) to produce a reasonably quantitative picture of the eigenfunction of each electron.

9.6 SELF-CONSISTENT FIELD CALCULATION OF EIGENFUNCTIONS

Slater orbitals are approximations based on empirical assumptions. However, they can be used as a starting point for a quantitative solution of the real problem mentioned in Section 7.1, the solution of the full Schrödinger equation that describes a many-electron atom.

Once again, it is necessary to make a model of the atom which enables solutions to be achieved with minimal calculation. In this case the model is the same as in the previous section. A single electron is seen as moving in a potential energy field made up of two components, one due to the positive nucleus and a second which is a spherical average of the potential due to all the other electrons. Then the Schrödinger equation can be integrated numerically to give an expression for the electron eigenfunction.

However, this presupposes that the eigenfunctions of all the other electrons are already known. To get round this difficulty, the method of **self-consistent fields** is employed. If a first approximation to the eigenstates of the other electrons is assumed (for example the Slater orbitals), the potential field in which the electron of interest moves can be calculated, and hence the Schrödinger equation can be solved to give an expression for that electron's eigenfunction. This expression can be used in a new calculation for the potential energy field in which a second electron moves and the Schrödinger equation can be solved for this electron. The process can be repeated until new expressions have been produced for the eigenfunctions of all the electrons. This gives a second-order approximation.

Now, if the first electron is again considered, the potential field in which it moves can be recalculated using the new values of eigenfunctions for the other electrons and a third order of approximation can be made. This process can be repeated until the expressions for the eigenfunctions do not change within the limits of the precision of calculation. The fields produced by this set of eigenfunctions are then said to be 'self-consistent', and the eigenfunctions are taken to be good solutions of the Schrödinger equation for the system. This method was originally pioneered by Hartree. However Hartree's approach only includes the classical effects of the interaction between an electron and a classical 'smeared out charge cloud' due to the other electrons. It was pointed out by Fock that there was also a non-classical 'exchange' interaction. Including this effect makes a difference of some 10%–20% to the solutions worked out by the Hartree method. Solutions produced by the full treatment are therefore known as Hartree–Fock self-consistent field atomic orbitals (or HF-SCF-AOs for short).

The equations used in the calculation are relatively easy to understand and give a useful demonstration of the use of the Schrödinger equation and the extension of the solution for the helium atom, outlined in Chapter 8. The discussion is limited here to the spatially varying part of the state function for a particular electron. As in the case of helium (Appendix G), the eigenstate for a particular electron has a space-dependent and a spin-dependent part and the eigenstate of the whole system will be an antisymmetric linear combination of single electron eigenstates.

The Hamiltonian (energy) operator for the whole atom is given by

$$\mathrm{H} = \sum_i \mathrm{H}_i + {}^1\!/_2 \sum_i \sum_j (e^2/4\pi\varepsilon_o r_{ij}). \tag{9.2}$$

H_i is merely the hydrogen-like Hamiltonian for an electron in the field of the nucleus of charge Ze. The second term gives the potential energy of each electron due to all the other electrons (the double sum is over all values of i and j from 1 to Z, except terms where $i = j$).

Considering electron 1 with eigenfunction $u_i(1)$. It is possible to write the Hartree–Fock equation

$$[\mathrm{H}_i + \sum_j (2\mathrm{J}_j - \mathrm{K}_j)]\, u_i(1) = E_i u_i(1). \tag{9.3}$$

The summation is over j from 2 to Z and J_j is the Coulomb operator, defined by

$$\mathrm{J}_j\, u_i(1) = [\, \int u_j^*(2) e^2/4\pi\varepsilon_o r_{12}\, u_j(2)\, \mathrm{dv}\,]\, u_i(1).$$

As usual, the integral is taken over all space and it can be seen that this equation merely defines the average Coulomb potential energy of electron 1 in the field of electron 2, which has eigenfunction $u_j^*(2)$.

The non-classical interaction effect is defined via K_j, the exchange operator, defined by

$$\mathrm{K}_j\, u_i(1) = [\, \int u_j^*(2) e^2/4\pi\varepsilon_o r_{12}\, u_i(2)\, \mathrm{dv}\,]\, u_i(1).$$

When the final, self-consistent form of the eigenfunctions has been established, it is possible to use (9.2) to work out E_i by multiplying both sides by $u_i^*(1)$ and integrating over all space. When this is done, it is simple to consider the special case where there are only two electrons and return to the expressions already worked out for the energy of electrons in the helium atom (8.6).

In general the results confirm the simple picture of 'hydrogen-like' eigenfunctions. However, this approach is still open to a number of objections. First, as with all SCF calculations, the fact that a self-consistent solution has been reached does not necessarily mean that it is the only solution.

Second, and more serious, the calculation still rests on the picture of a single electron moving in the 'smeared out' spherically symmetric potential field due to the other electrons. Use of modern computers enables more sophisticated methods of solving the Schrödinger equation for the atom, but these are well beyond the scope of this book.

One final conclusion can be drawn from the discussion. The calculated 'orbitals' can be understood most clearly in terms of a continuous 'charge cloud' surrounding the nucleus. When all electrons are considered the atom can be modelled as a nucleus surrounded by a continuous cloud of charge and the shape of this charge cloud can be worked out from the spatial probability density for each electron given by the square modulus of the eigenfunction of the electron.

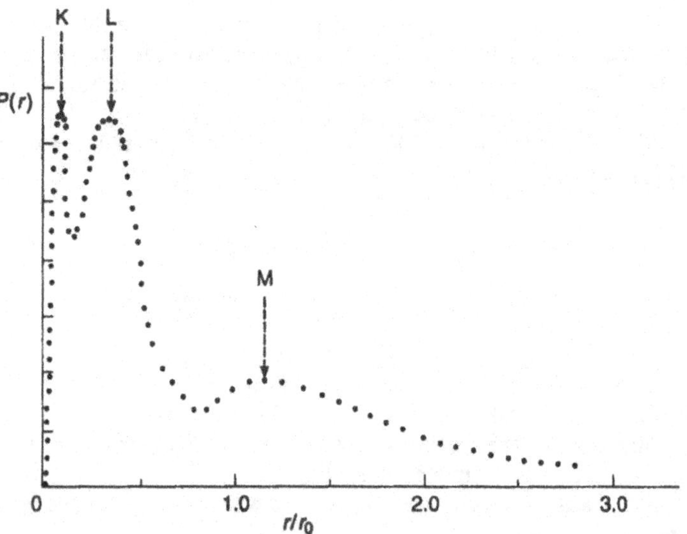

Fig 9.6 Picture of the radial distribution of the electron 'charge density' distribution in the ground state of argon, calculated by the Hartree method. The maxima K, L, and M correspond to the 'filled shells' given by the Aufbau picture.

An example of Hartree's calculation for argon is given in Fig. 9.6 This shows clear maxima equivalent to the particular 'closed shells' discussed earlier. Solutions of the HF-SCF-AO calculations are available, usually in tabular form, but are quite difficult to understand in terms of physical pictures.

9.7 THE EFFECTS OF SPIN–ORBIT COUPLING ENERGY – COUPLING SCHEMES

The eigenfunctions calculated by the methods described in Section 9.6 do not, of course, include the effects of the spin–orbit coupling energy discussed in Chapter 5, as this energy is relatively small when compared to Coulomb and exchange energies. However, when discussing fine structure and term series, it is necessary to go into more detail than that given in Chapter 8. Although in many cases (particularly for light elements) the term symbol and Hund's rules give a good basis for discussion, for many elements they do not.

The problem arises from the fact that in a multi-electron atom, there are a number of independent sources of angular momentum as each electron has its own orbital and spin angular momenta. Classically, this presents no problem as angular momentum vectors may be added to give a unique resultant angular momentum for the whole system. This is not true for quantized angular momenta, as can be seen from Appendix B.

For example, consider an atom which comprises a number of closed shells with two electrons outside them. The four electron angular momentum operators may be combined in two ways. The two orbital angular momentum operators $\mathbf{l}_1$ and $\mathbf{l}_2$ may be combined to give a resultant orbital angular momentum operator $\mathbf{L}$ with well-defined eigenvalues for its amplitude and z-component. Similarly the two spin operators $\mathbf{s}_1$ and $\mathbf{s}_2$ may be combined to give a resultant spin operator $\mathbf{S}$ with well-defined eigenvalues for its amplitude and z-component. Finally, $\mathbf{L}$ and $\mathbf{S}$ may be combined to provide a resultant operator $\mathbf{J}$ which will have a set of well-defined values for its amplitude and z-component. As shown in Section B.5, these values can be written down in terms of the quantum numbers defining eigenvalues of $\mathbf{L}$ and $\mathbf{S}$.

However, there is another way of doing the addition. Suppose the first step is for each electron to combine its spin and orbital angular momentum operators to give resultant angular momentum operators $\mathbf{j}_1$ and $\mathbf{j}_2$. The next step would then be to combine $\mathbf{j}_1$ and $\mathbf{j}_2$ to give $\mathbf{J}$. In this case, the values of $\mathbf{L}$ and $\mathbf{S}$ cannot be defined. In both cases, there must be the same resultant set of values of $\mathbf{J}$, since it is merely the method of arriving at the end point that has differed. The first method is normally known as the LS method and the second as the jj method.

The two methods correspond to two different physical situations. The LS case is appropriate when the energy due to spin–orbit coupling for an individual electron is small compared with the energy due to the electric repulsion between the two electrons. In the jj case, the spin–orbit energy is large compared with the repulsion energy. The difference is illustrated in Fig. 9.7, for the pair of electrons discussed above, assumed to have configuration (*npns*). On the left, the LS coupling scheme shows the ^1P and ^3P states separated in energy by the large effect due to the exchange interaction (Hund's first rule) and the small spin–orbit energy merely serving to separate the triplet levels.

On the right, the four states are defined in terms of the possible values of j_1 and j_2 and the possible resultant values of J. In this case, these levels are separated out by the small electron–electron coupling. However, it is still possible to identify unambiguously each state in the jj coupling scheme with one in the LS coupling scheme, so that each state may be defined by a term symbol.

However such a definition is not really useful. First, the relative energies of states in a particular configuration cannot be deduced from Hund's rules. Second, and more important, the selection rules for allowed transitions, $\Delta S = 0$ and $\Delta L = \pm 1$, no longer hold.

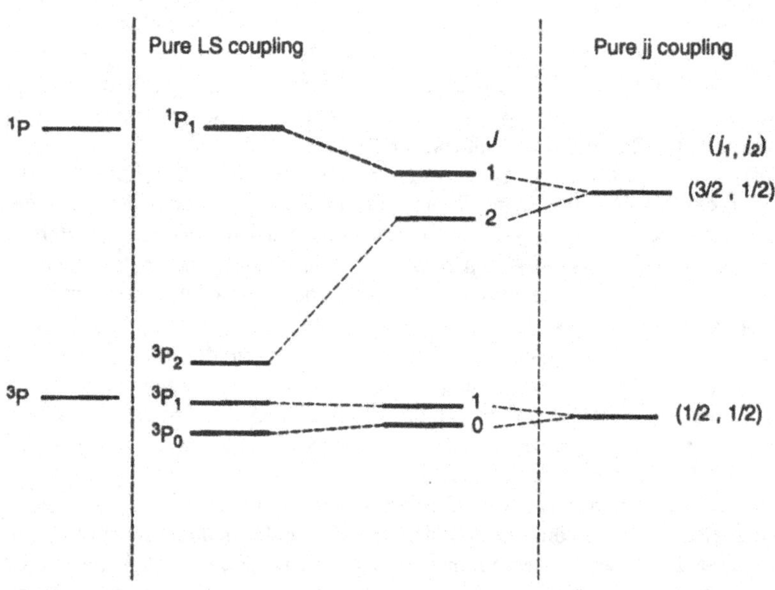

Fig. 9.7 Demonstration of the various coupling schemes for a system consisting of one s and one p electron. Starting on the left, levels are separated into singlet and triplet terms. For LS coupling, the fine structure splitting is as shown. As the electron–electron interaction energy becomes relatively smaller compared with the interaction between each individual electron and the nucleus, the splitting changes until the pure jj coupling state on the right hand side obtains.

LS coupling is the most common form of coupling encountered and is normally known by the name of 'Russell–Saunders coupling' after the physicists who first investigated it. However there are many important cases in which it does not apply, especially in the case of heavier elements. As might be expected, even in a particular group of the periodic table, the coupling scheme varies with atomic number. For example, in group IV of the periodic table, carbon and silicon exhibit practically pure LS coupling. However, moving through the group, germanium, tin and lead approach closer and closer to pure jj coupling.

To make things even more complicated, it can be seen that even in a particular atom different configurations may be closer to one form or another of the pure coupling schemes. Although ground states may exhibit LS coupling, excited states may well tend towards jj. An example is lead, where although the excited states must be described through jj coupling, the ground state ($^3P_0$) is best described in terms of Russell–Saunders coupling, even though the energy splitting of the three triplet terms is very large.

In general, as Z increases the excited states move first towards jj coupling since an electron with a large value of the principal quantum number n is rather weakly coupled to the other electrons. For example in excited states of neon (which have an importance in laser systems) the inner electrons will show Russell–Saunders coupling, producing a resultant J_c which is then coupled to the j of the outer electron.

For other elements there are very many complex and transitional coupling schemes which cannot be discussed here.

10

The atomic physics of lasers

In the first fifty years of atomic physics the vast majority of experimental investigations involved the observation of the spontaneous emission of radiation from atoms in gas discharges. It was only in the late 1950s that the importance of stimulated emission was realized.

10.1 STIMULATED EMISSION AND AMPLIFICATION

As discussed in Chapter 6, the vital importance of stimulated emission is that light stimulated by a radiation field is emitted in phase with that field and therefore adds to its amplitude in a coherent way. Because of this, it is straightforward to work out an expression for the amplification of a beam of electromagnetic radiation travelling through a tube of gas atoms. As before, two

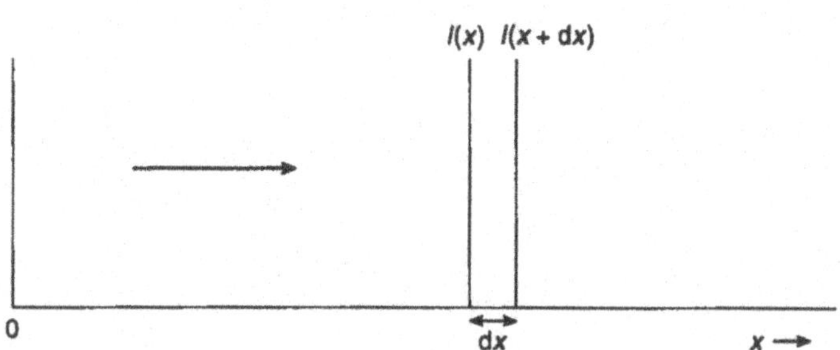

Fig. 10.1 Amplification or absorption in a tube of gas. A plane wave travels from the left. Intensity at the origin is $I(0)$. Change dI over distance dx is proportional to $I(x)$.

levels are considered with atomic population densities N_1 and N_2 ($E_2 > E_1$). A beam of radiation with energy density $\rho(\nu_{12})$ travels down the tube (of cross-section area $S\,m^2$) in the x-direction. It is assumed that the intensity of the beam is uniform across the tube cross-section (Fig. 10.1).

Considering a section of the tube of thickness dx at distance x from the origin, the number of quanta of radiation added to the field in time dt is given by

$$dn/dt = (N_2\text{-}N_1)\, B_{12}\, \rho(\nu_{12})\, Sdx. \tag{10.1}$$

Now, $\rho(\nu_{12}) = I(x)/c$, where $I(x)$ is the radiation intensity at x, and the rate of increase in radiation energy in the volume Sdx

$$d\rho/dt = h\nu_{12}\, dn/dt = S\, dI.$$

Therefore, substituting and using the definition of radiation intensity

$$dI = [(N_2\text{-}N_1)\, B_{12}\, h\nu_{12}/c]\, I(x)\, dx. \tag{10.2}$$

There are two situations. If $N_2 < N_1$, radiation travelling along the tube suffers absorption. On the other hand, if $N_2 > N_1$ the radiation is amplified. This is the basis of laser action.

In either situation, the solution of (10.2) gives an expression for the intensity in the cavity as a function of x

$$I(x) = I(0)\exp(\alpha x) \tag{10.3}$$

where

$$\alpha = [(N_2\text{-}N_1)\, B_{12}\, h\nu_{12}/c]. \tag{10.4}$$

When $N_2 > N_1$, $I(x)$ increases exponentially. In this case, α is known as the small-signal gain per unit length with units of m^{-1} (it is sometimes written as 'percent per metre').

It is interesting to speculate why the possibility of using this mechanism to build a light amplifier went unnoticed for fifty years after Einstein's original work. Basically, amplification only takes place if $N_2 > N_1$. For atoms in thermal equilibrium, the ratio N_2 / N_1 is given by Boltzmann's law and is of the order of 10^{-5} if ν_{12} is in the visible region of the spectrum. Physicists tended to assume that atomic systems were invariably in thermal equilibrium, so that in the experimental situation outlined above any stimulated emission would be completely swamped by absorption. However, in the 1950s devices using stimulated emission to produce oscillation in the microwave region were invented (masers). These used the vibrational energy levels of ammonia molecules and relied on the fact that molecules in a higher vibrational state could be separated from those in a lower state by a simple electrostatic arrangement.

Only in the 1960s was it realized that the thermal equilibrium population distribution between optically-connected atomic energy levels was not inevitable. To emphasize what was assumed to be the unusual nature of this occurrence, situations where $N_2 > N_1$ were called situations of 'population inversion'.

However as soon as the taboo was broken, many ways of producing such population inversions were swiftly developed.

10.2 THE LASER OSCILLATOR

Even assuming that the condition $N_2 > N_1$ can be achieved, (10.3) shows that α is quite small for normal gas densities. Thus for perceptible amplification to take place, very long tubes of gas would be required.

This problem is overcome by reflecting the radiation back on itself so that it travels many times through the same relatively short tube of gas. The device thus produced is known as a laser (an acronym for Light Amplification by Stimulated Emission of Radiation). In essence, the device involves a tube of gas placed between two parallel highly reflecting mirrors (Fig. 10.2). The amplification process then becomes part of a more complex system where the pair of mirrors functions as a resonant cavity. Light travelling between the mirrors interferes and constructive interference only takes place if the following resonance

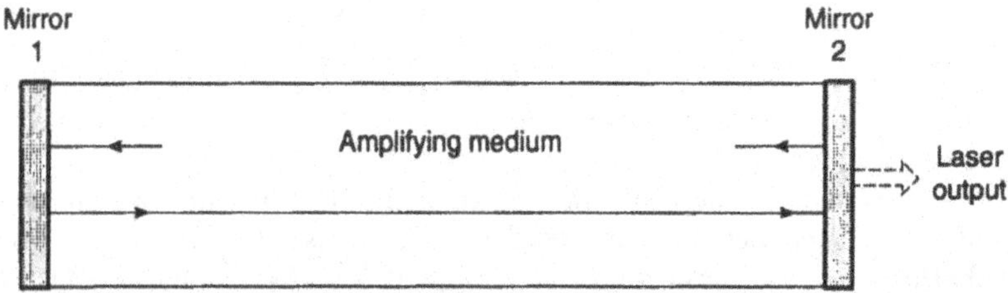

Fig. 10.2 Light amplification in a laser cavity. A wavefront bounces back and forth between the parallel mirrors. Mirror 2 is not perfectly reflecting, so a small part of the wave front is transmitted through it, providing the laser output.

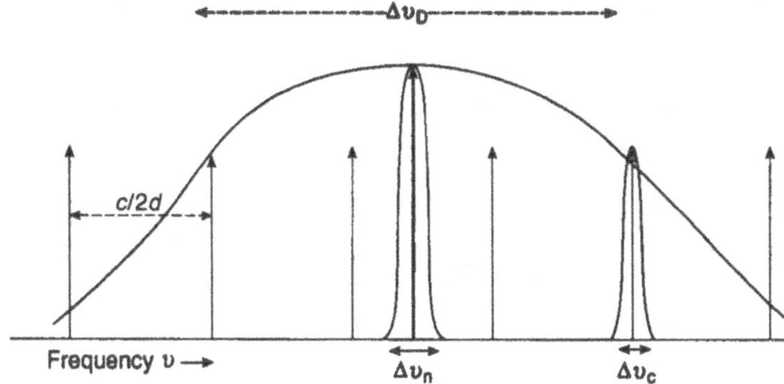

Fig. 10.3 Cavity resonance frequencies and Doppler broadened spectral line. Each cavity resonance has the 'passive' width $\Delta\nu_c$. $\Delta\nu_n$ is the natural linewidth of the transition.

relationship is satisfied

$$\nu = n\ c/2d \tag{10.5}$$

where d is the distance between the mirrors and n is an integer.

Assuming that this condition is satisfied, a standing-wave field is established in the space between the mirrors. The gas amplifier now acts like the amplifier in a conventional electronic oscillator. The restriction on the values of possible resonant frequencies, (10.5), might appear to be very restrictive. However, putting values into the equation shows that resonant frequencies are closely spaced, with a spacing much smaller than the Doppler line width of the typical atomic transition (Fig. 10.3). The natural linewidth of the transition is also not much smaller than the mode spacing. The standing-wave field may be considered as the sum of two travelling-wave fields travelling in opposite directions in the laser cavity. Each of these waves interacts with the atoms whose relative speeds produce a Doppler shift in the transition which coincides with the frequency of the mode (or, to be more precise, those atoms where the cavity mode falls within their natural linewidth). Each travelling wave interacts with a separate set of atoms, but, as far as the cavity resonance is concerned, the gas functions like a broad-band amplifier (Fig. 10.4).

Mirrors never have perfect reflectivity, so energy is lost in the reflection process. Energy is also lost by diffraction of the wave round the edges of the mirror, but it may be shown that this loss is small compared to reflection losses if the mirrors are accurately aligned. Assuming that the only energy losses in the cavity are due to the reflectivity R of each mirror, (10.3) and (10.4) may be used to calculate the value of $(N_2\text{-}N_1)$ which will enable a standing-wave field to be established in the cavity.

Considering a wavefront that starts at some point near the first mirror, travels through the gas, suffers reflection at the second mirror, travels back through the gas, suffers reflection again at the first mirror and returns to its start point, an expression may easily be written down for the intensity of the standing wave to be sustained:

$$I(2d)/I(0) = R^2 \exp(2\alpha d) \geq 1. \tag{10.6}$$

Therefore for the field intensity in the cavity to be self-sustaining,

$$2d\,(N_2 - N_1)B_{12}\,h\nu_{12}/c \geq \log(1/R^2). \tag{10.7}$$

Equation (10.7) defines the 'threshold condition' for laser operation.

It should be noted that (10.6) implies that once the threshold condition is exceeded, the intensity of the field in the cavity will continue to increase exponentially. This will not, of course, continue indefinitely. Saturation will occur. As will be seen shortly, some external mechanism is required to produce a value of (N_2-N_1) greater than the threshold value. As the field intensity in the cavity increases, the mechanism of stimulated emission itself will act to reduce the value of (N_2-N_1). At a certain value of I, stimulated emission will be

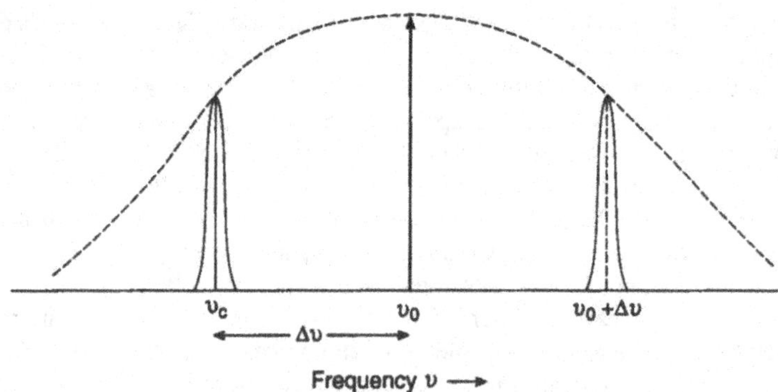

Fig. 10.4 Interaction between a gas of atoms and a cavity mode. The cavity resonance frequency, ν_c is $\Delta\nu$ lower than the atomic transition frequency ν_0. The cavity mode standing wave is made up of two travelling waves. One interacts with atoms moving so that their transition frequencies are Doppler shifted close to $\nu_0 - \Delta\nu$, the other with atoms moving so that their transition frequencies are shifted close to $\nu_0 + \Delta\nu$. Thus gain is produced by the two sets of atoms shown in the diagram. When $\nu_c = \nu_0$, there is only one set of atoms to interact with, so the gain is less. This is known as the 'Lamb dip'.

removing so many atoms from the higher state that the threshold condition is only just maintained. This is the intensity I_{sat}, at which saturation occurs (see Fig. 10.5).

The above analysis gives a brief qualitative picture of the physics of gas laser operation and gives enough detail to enable a discussion of the atomic physics involved in producing a population inversion. It should be noted that the actual 'laser radiation' is the part of the field which is lost to the laser cavity via the non-perfect reflectivity of the mirror (in normal systems, one mirror is made as highly reflecting as possible, while the laser radiation is taken from the other). Obviously for a given laser, with an amplifying medium of given gain, there is an optimum value for the transmittivity of the mirror to give maximum output.

One final point worth discussing is the linewidth of the laser radiation. This is primarily a function of the properties of the resonant cavity, and not of the amplifying system. This may seem surprising, but a calculation of the width of the resonance in a 'passive' cavity with no amplifying medium present, gives a value of around 10^8 Hz. The positive feedback of the amplifying system 'sharpens up' this width, just as in the case of electronic oscillators.

The linewidth of the the laser radiation can be estimated from the general definition of the 'Q' of a resonant circuit

$$\begin{aligned} Q &= \nu_l / \Delta\nu_l \\ &= 2\pi \text{ (energy stored in the oscillator)/ energy lost per cycle.} \end{aligned} \tag{10.8}$$

Given the laser cavity shown in Fig. 10.2, this boils down to

$$Q = \nu_l / \Delta\nu_c = 2\pi\, d\, \nu_l / c(1 - R). \tag{10.9}$$

The reflectivity R may be of the order of 95%, giving a typical value for the width of the passive cavity resonance of 10^8 Hz. Calculation of an expression for the laser linewidth is well beyond the level of this book, but it is possible to quote a simple expression for the linewidth

$$\Delta\nu_l = (n + 1/2)\, h\nu_l\, \Delta\nu_c / W \tag{10.10}$$

where n is the mean number of 'noise' quanta in the mode (due to spontaneous emission, thermal energy etc.) and W is the mean laser energy in the mode. Typically, $n \approx 1$, giving a value of $\Delta\nu_l$ of the order of 10^{-1} Hz.

Such a linewidth could only be realized if the distance d between the mirrors were held stable to within better than 1 part in 10^{15}. In practice, the value of the laser linewidth is usually greater than 10^3 Hz, mainly due to thermal fluctuations in the optical length of the cavity, but it can be seen that this linewidth is dramatically smaller than the natural linewidth of radiation produced by spontaneous emission.

10.3 SEMI-CLASSICAL THEORY OF THE LASER

The picture of laser operation outlined in the preceding sections gives a sufficient understanding for most general purposes. However, the more detailed semi-classical theory of the laser developed by Lamb in the early 1960s provides a fascinating example of the use of the concepts of field–atom interactions developed in Chapter 6. It also enables nearly all the characteristics of the laser to be discussed both qualitatively and quantitatively.

In this case the interaction is between a system of atoms and a standing wave field and involves a 'self-consistent' approach, though not an iterative successive approximation one like the Hartree–Fock calculations discussed in Chapter 9.

The simple model of the laser cavity shown in Fig. 10.2 is considered here, although the theory can be used for three-mirror or more complicated cavities. The axis of the cavity is taken as the z-direction and the electromagnetic field in the cavity is assumed transverse, but constant in the (x, y) plane. For simplicity, the vector nature of the field is not considered in this case.

In a perfect cavity, with no losses at the mirrors or in the medium between, the electric field $E(z,t)$ can be written as a superposition of standing wave fields (the normal modes of the field in the cavity)

$$E(z,t) = \Sigma\, E_n\, U_n(z) \cos(2\pi\, \Omega_n t) \tag{10.11}$$

where E_n is a constant and Ω_n is the cavity resonant frequency, given as before by

$$\Omega_n = nc/2d \text{ Hz.}$$

The function $U_n(z)$ defines the variation in amplitude of the mode along the axis of the cavity

$$U_n(z) = \sin(n\, \pi\, z/d).$$

Equation (10.11) is, of course, merely the solution of the loss-free version of Maxwell's wave equation, with appropriate boundary conditions.

In a laser cavity, there are real losses at the mirrors which are balanced by the gain produced by the laser medium. Thus the perfect cavity modes of (10.11) no longer apply. However, to a good approximation, the sum over modes can be replaced by a sum over 'quasimodes'

$$E(z,t) = \Sigma\, E_n(t)\, U_n(z) \cos(2\pi\, \nu_n t + \phi_n(t)). \tag{10.12}$$

The terms $E_n(t)$ and $\phi_n(t)$ are now time-dependent amplitude and phase functions which vary slowly compared with times of the order of the period of the radiation. The frequency ν_n is the frequency of laser oscillation, subject to the slow variation of $\phi_n(t)$.

To produce this result, two additional terms are included in the classical wave equation, a loss term σ and a driving polarization term P. Assuming that the cavity losses are not frequency-dependent, σ is simply defined in terms of the cavity Q defined in (10.8), as

$$\sigma = \nu/2Q$$

where ν is a frequency close to those of interest (Ω_n and ν_n).

More interesting is the polarization P which drives the field and is the result of the interaction between the amplifying medium and the cavity field. This term can be decomposed in the same quasimodes as the field, but is not necessarily in phase with it. It can be written as

$$P = \Sigma\left[C_n(t)U_n(z)\cos(2\pi\nu_n t + \phi_n(t)) + S_n(t)U_n(z)\sin(2\pi\nu_n t + \phi_n(t))\right] \quad (10.13)$$

where $C_n(t)$ and $S_n(t)$ are the in-phase and quadrature parts of P.

Substituting back into the wave equation, using orthogonality, neglecting terms in derivatives of $E_n(t)$ etc., and separating out sine and cosine coefficients enables a pair of remarkably simple equations to be written down for each quasimode

$$dE_n(t)/dt = -(\nu/2Q)E_n(t) - (\nu/2)S_n(t) \quad (10.14a)$$

$$\nu_n + d\phi_n(t)/dt = \Omega_n - (\nu/2)C_n(t)/E_n(t). \quad (10.14b)$$

These equations define the characteristics of the modes. Equation (10.14a) defines the time-variation of the amplitude $E_n(t)$ of the laser mode, while (10.14b) defines the laser frequency ν_n.

The important point, and the kernel of the semi-classical analysis is that the terms S_n and C_n are the result of the interaction of the field on the amplifying medium. S_n and C_n are functions of E_n, so (10.14a) and (10.14b) are, in effect, self-consistency equations for the field in the cavity.

It now remains to work out expressions for S_n and C_n, using the time-dependent perturbation approach described in Sections 6.4 and 6.5. The polarization function must include a form of summation over the large number of individual independent atoms that interact with the field. If the atoms are stationary, this presents no problem, as the $U_n(z)$ functions defined in (10.11) allow for the varying field amplitude along the laser axis. However, in a gas laser, the motion of gas atoms is such that an atom may well move between points of maximum electric intensity and points of minimum intensity in times of the order of the rates of change of $E_n(t)$, which makes for extremely complex calculations.

For the present work, it is only necessary to present the important physical results, rather than the detailed analysis. The analysis of the atom–field interaction in Chapter 6 gives the conclusion that to a first order of approximation, S_n and C_n must be linear functions of E_n and $(N_2\text{-}N_1)$, where N_1 and N_2 are the numbers of atoms per unit volume in the lower and upper states which interact with the field mode. Considering first the laser amplitude, (10.14a) may be rewritten

$$dE_n(t)\,/dt = [-\,(\nu/2Q) + K(N_2\text{-}N_1)\,]\,E_n(t) \tag{10.15}$$

where K is a constant which involves the dipole moment matrix element for the transition.

This equation predicts an exponential growth for the mode amplitude E_n (note that this is a growth in time of a standing wave mode, not the growth in distance of a travelling wave discussed in Section 10.1). For saturation to occur, the perturbation analysis must be taken to a third order, giving a further term dependent on E_n^3. Equation (10.15) may then be rewritten simply in the form

$$dE_n\,/dt = \alpha_n\,E_n - \beta_n\,E_n^3. \tag{10.16}$$

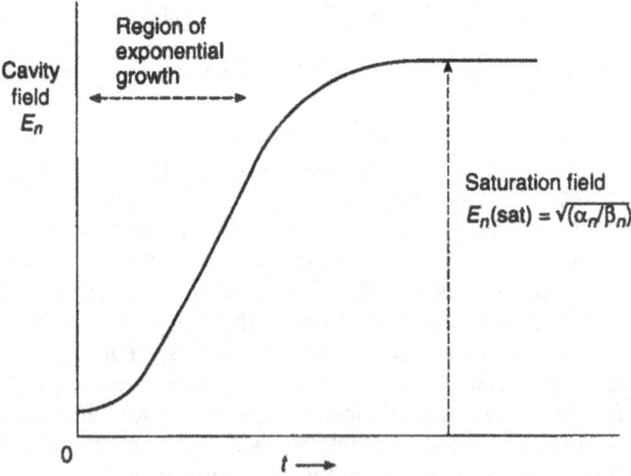

Fig. 10.5 Solution of (10.16) for positive values of α_n. Note that $E_n(0)$ must be non-zero. This is a consequence of the semi-classical approach.

The solution of this for positive values of α_n is shown in Fig. (10.5). It gives an exponential growth to saturation, with α_n defining the small-signal gain parameter and β_n defining a saturation parameter. The approximation rests on the value of α_n, and hence the saturated value of E_n, not being too great.

An equivalent analysis for (10.14b) gives an expression for the steady state laser frequency

$$\nu_n = \Omega_n + \sigma_n - \rho_n E_n^2. \tag{10.17}$$

In this case, the laser frequency is 'pulled' from the passive cavity frequency by the loss term σ_n and 'pushed' towards it by the intensity-dependent term, $\rho_n E_n^2$.

10.4 MODE INTERACTIONS

A major strength of the semi-classical theory is that it gives a quantitative picture of the interaction between laser modes discussed in Section 10.2.

In the equations derived in Section 10.3, it has been assumed that each laser mode interacts independently with a population of atoms. For stationary atoms, if the mode spacing is large compared with the natural linewidth of the transition, only a single mode will be excited, and its amplitude will be maximum if the laser frequency ν_n coincides with the transition frequency ν_0. However, as described qualitatively in Section 10.2, in a gas laser the modes interact with a population of atoms which have their transition frequencies shifted by the Doppler effect. This interaction can be described quantitatively by the semi-classical theory. Simply, the constant K in (10.15) now contains an exponential of the form given in (5.14), so that several modes may have positive values of α. These modes will all be excited, and there will be interaction terms linking them.

For example, suppose that two modes are excited. In this case, the amplitude-defining equation (10.16) is developed into a pair of coupled equations

$$\begin{aligned} dE_1/dt &= \alpha_1 E_1 - \beta_1 E_1^3 - \Theta_{12} E_1 E_2^2 \\ dE_2/dt &= \alpha_2 E_2 - \beta_2 E_2^3 - \Theta_{21} E_1^2 E_2 \end{aligned} \tag{10.18}$$

where modes (labelled 1 and 2) are linked by coupling coefficients Θ_{12} and Θ_{21}.

The solutions of this pair of equations can be understood easily in terms of the discussion in Section 10.2. If the two modes with frequencies ν_1 and ν_2 are positioned asymmetrically with respect to the natural frequency of the transition ν_0, the modes do not interact with the same sets of atoms and coupling is weak. Both modes are excited, stable values of E_1 and E_2 can be calculated from

(10.18) and the laser output will consist of two frequency components. If, on the other hand, the two modes are symmetrical with respect to ν_0, strong coupling will occur and one mode may 'quench' the other.

Similar, but more complex equations may be written down where three or more modes are in question. On the other hand, if the length of the laser cavity is such that only one mode can be excited, the solution of (10.16) in terms of $(\nu_n - \nu_0)$ exhibits a minimum for the value $(\nu_n - \nu_0) = 0$ where the mode only interacts with one set of atoms. This effect is observed experimentally and is known as the 'Lamb dip' (see Fig. 10.4)

Finally, the same coupling effect occurs in the frequency determining equations. Once again, for two mode operation, a pair of coupled equations occur, defining the steady state laser frequencies

$$\begin{aligned} \nu_1 &= \Omega_1 \cdot \sigma_1 - \rho_1 E_1^2 + \tau_{12} E_2^2 \\ \nu_2 &= \Omega_2 \cdot \sigma_2 - \rho_2 E_2^2 + \tau_{21} E_1^2. \end{aligned} \tag{10.19}$$

Here a further interesting phenomenon occurs. Returning to the original definition of the quasimode, the term ϕ_n specifies the phase variation of the nth mode.

For two or more modes, a 'relative phase angle', Φ, between the modes can be specified in terms of linear combinations of the various ϕ_n. In the two mode case, Φ is given by the equation

$$\Phi = (\nu_2 - \nu_1)t + \phi_2 - \phi_1. \tag{10.20}$$

In cases of weak coupling, it can be shown that such relative phase angles vary randomly in time, so that the output of a continuous laser consists of a number of unconnected monochromatic beams of radiation. However, for certain values of coupling, the relative phases become 'locked' together. In this 'mode-locked' regime, the various frequency components in the output add together coherently to give an output consisting of a continuous train of pulses, separated by a time period $2d/c$ s.

10.5 QUANTIZED-FIELD LASER THEORY

Semi-classical theory enables a very large number of laser phenomena to be described quantitatively. As can be seen from the discussion of Section 6.9, this follows from the fact that a single mode of the laser cavity is excited to a very high photon number and is best described quantum mechanically by a coherent state which, for high photon numbers, gives a good approximation to a classical field with well-defined $\boldsymbol{E}$ and $\boldsymbol{B}$ vectors.

However, just as in the cases discussed in Chapter 6, there are certain areas that cannot be discussed classically. For example, a return to (10.16) shows that it is necessary for there to be a non-zero value of E at time $t = 0$ for the equation to lead to a steady state solution with non-zero E_n. This problem is conceptually identical to the question of the existence of spontaneous emission, discussed earlier. In qualitative terms, it coincides with the statement that spontaneous emission in the cavity is required to 'start the mode off'.

Similarly, the ultimate linewidth of laser radiation can only be discussed satisfactorily in terms of the 'statistics' of laser photons. Such considerations generate interesting new phenomena, such as the production of 'squeezed states' of light, which cannot be outlined here.

10.6 PRACTICAL ATOMIC GAS LASERS

In order to work out how to obtain the population inversion necessary for laser action, it is possible to use the rate equation approach discussed in Chapter 6. Assuming that some mechanism exists which populates level 2 at a constant rate R_2 atoms per second and level 1 at rate R_1, rate equations can be written down to describe all processes taking place (Fig. 10.6)

$$\begin{aligned} dN_2/dt &= R_2 - A_{21}N_2 - (N_2 - N_1)\,B_{12}\,\rho \\ dN_1/dt &= R_1 + A_{21}N_2 - A_{1x}N_1 + (N_2 - N_1)\,B_{12}\,\rho. \end{aligned} \tag{10.21}$$

Here A_{1x} describes processes which depopulate level 1 by spontaneous emission

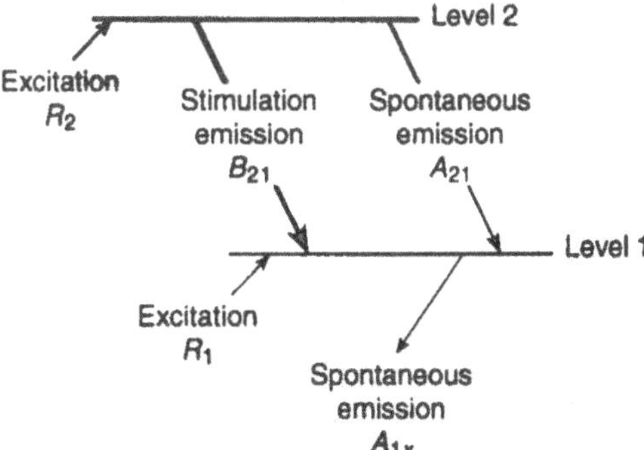

Fig. 10.6 Excitation and de-excitation processes for two laser levels. (Absorption from level 1 to level 2 has been omitted for clarity.)

to other levels (it is assumed that there are no other transitions from level 2). Unit volume of the gas is considered and ρ is the energy density of the mode of the cavity field which is interacting with the gas.

A similar rate equation can be written for the energy in the electromagnetic field in the cavity mode. This equation derives from a simplification of the picture given in Section 10.3 but is quite adequate for the present discussion

$$d\rho/dt = h\nu(N_2 - N_1)\, B_{12}\, \rho - \kappa\rho. \tag{10.22}$$

Here, the term κ is used to define the loss of energy per unit time through the cavity mirrors (other sources of loss being assumed small). From the definition of the Q of the cavity (10.9)

$$\kappa = c(1 - R)/\, d.$$

Assuming that the population inversion mechanism is switched on at time $t = 0$ and the energy density in the cavity is essentially zero, a steady-state solution for (10.21) gives

$$N_2(0) = R_2\, /A_{21};\quad N_1(0) = (R_1 + R_2)/A_{1x}. \tag{10.23}$$

This shows that two criteria are required for population inversion: the rate of population of the upper level must be greater than that of the lower level and the rate of depopulation of the lower level by spontaneous emission must be larger than that of the upper level. If the values of R_1, R_2, A_{21} and A_{1x} are known, (10.21) can be used to calculate a value for the population inversion. If this is greater than the threshold population inversion defined by (10.22), laser energy will build up in the cavity. Equations (10.21) also show how this build-up of energy will decrease the population inversion until it falls to the threshold value. At that point the system will be in equilibrium, with $dN_1/dt = dN_2/dt = 0$. Equations (10.21) and (10.22) can be solved to calculate the steady-state laser intensity.

At equilibrium, the population inversion is given by

$$(N_2 - N_1) = [R_2(1/A_{21} - 1/A_{1x}) - R_1(1/A_{1x})]/(1 + B_{12}\rho\, /A_{21}). \tag{10.24}$$

Equation (10.24) defines a maximum value of the energy density in the mode for which the population inversion is just great enough to add energy to the field at the same rate as it is being lost. This defines the saturation or steady-state field in the mode.

Equations (10.21) and (10.22) can be solved to describe the development of the field in the mode from zero to steady state. A typical result is shown in Fig. 10.7. Qualitatively, it can be seen from (10.22) that if the net gain in the cavity is small, the growth of the field will also be slow.

The model described above gives a highly simplified, but useful picture of the field–atom interaction. Three further points are worth noting. First, the interaction is with a single mode of the standing wave field in the cavity. This field will have nodes where the field (and hence the interaction) is minimum and antinodes where the interaction is maximum. Thus, the value of $(N_2 - N_1)$ will vary along the axis of the laser cavity between its zero-field and its saturation value with a periodicity equivalent to that of the field. Second, it has been assumed that the frequency of the field is exactly in resonance with the central frequency of the transition. In fact, as mentioned in section 10.3, line broadening (of all types) must be taken into consideration. The effect of Doppler broadening was mentioned earlier. Natural broadening also affects the strength of the interaction. In essence, the naturally broadened line acts like a tuned amplifier. When the cavity frequency coincides with the central frequency amplification is at a maximum, but 'off tune' the gain decreases in proportion to the Lorentzian

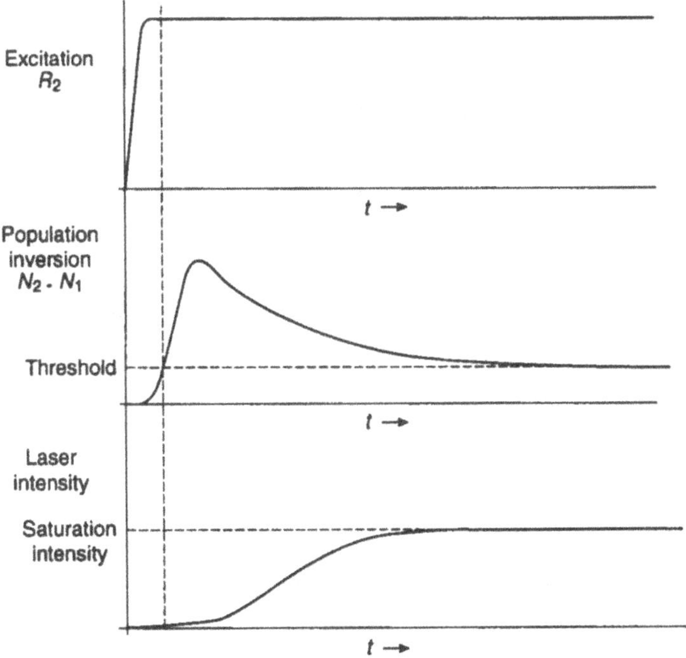

Fig 10.7 Time development of population inversion and radiation energy in a laser cavity. Depending on the various time constants, an 'underdamped', oscillatory approach to the steady state is also possible.

function, (5.15). Finally, as noted earlier, the absorption/stimulated emission process in (10.21) needs some field energy in the mode to start it off. This is assumed to come from spontaneous emission.

It would be nice to think that lasers based on atomic processes were invented using the above analysis. However, the various parameters are hardly ever known to enough precision to make such a detailed approach worthwhile. Instead, most successful laser systems were invented by a process of inspired guesswork or by trial and error.

10.7 THE ARGON ION LASER

A good example is the argon ion laser. Electron collisions in a tube of argon gas produce a spectrum due to excited states of argon. However, argon in the tube is also ionized (its ionization energy is 15.8 eV) and a spectrum due to states of the argon ion Ar^+ is also seen. The electron configuration of the ion is

$$(1s)^2(2s)^2(2p)^6(3s)^2(3p)^5$$

so that its energy level diagram is 'hydrogen-like', with the difference that for a given term, states of higher J lie lower than those of lower J. A simplified version of the energy level diagram is shown in Fig. 10.8.

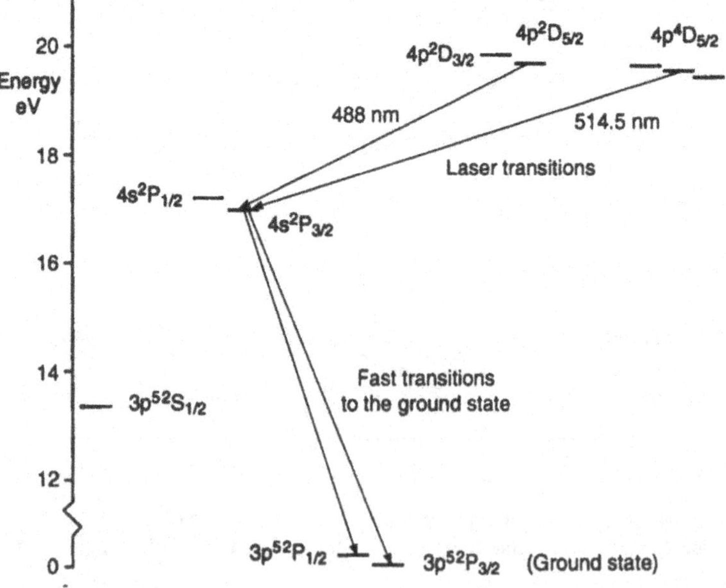

Fig. 10.8 Simplified energy level diagram for Ar^+. The most important laser transitions are shown. (Note that in this case energies are measured upward from the ground state.)

It turns out that transitions from the 4s ^2P states take place with a particularly short time constant (around 10^{-10} s), while those from higher states have a more normal time constant of 10^{-8} s. It is therefore found that a population inversion normally exists between higher-lying levels and this state. This is particularly marked for the two transitions

$$4p\ ^4D_{5/2} \rightarrow 4s\ ^2P_{3/2}$$

$$4p\ ^2D_{5/2} \rightarrow 4s\ ^2P_{3/2}.$$

These transitions may be used for amplification in a laser oscillator. The first produces light at a wavelength of 514.5 nm in the green region of the spectrum and the second at 488 nm in the blue region.

The argon ion laser is perhaps the most useful general purpose gas laser. Production of the population inversion merely involves the production of a discharge in a tube of argon gas. The more current there is in the discharge, the more ions are excited into the upper levels and hence the more intense the laser output.

10.8 THE HELIUM–NEON LASER

Perhaps more common than the argon ion laser is the helium–neon laser. The amplifying system in this case is more complicated and cannot produce such high population inversions, but it displays some nice points of atomic physics.

The starting point is the energy level diagram of helium (Fig. 8.2). It will be recalled that the two states 2s $^1S_0$ and 2s $^3S_1$ are metastable. This means that as a discharge in helium populates higher levels, atoms in some of those levels decay down to the 2s $^1S_0$ and 2s $^3S_1$ levels, but cannot decay further.

This fact is not, of course, any use directly for laser amplification. However it was noticed by Schawlow and Townes in 1961 that the energy gap between the metastable levels and the ground state is almost exactly the same as that between some higher-lying levels in neon and the neon ground state (Fig. 10.9). This means that in a mixture of helium and neon there is a high probability for a so-called 'collision of the second kind' – a collision between excited helium and ground state neon atoms, which will transfer the excitation energy from the helium atom to the neon atom. This process will selectively populate some high-lying levels in neon and a population inversion will be produced between these levels and levels of lower energy which do not gain in this way.

The physics of the process is quite complicated. This is exacerbated by the fact that the neon levels are not easily describable via the term symbols defined in Chapter 8 (see section 9.7). Because of this, the neon levels in Fig. 10.9 are merely described in terms of their electron configurations. The important transitions are between levels in the configurations

$(2p)^5 4s \rightarrow (2p)^5 3p$ at a wavelength of 1152 nm
$(2p)^5 5s \rightarrow (2p)^5 3p$ at .. 633 nm
$(2p)^5 5s \rightarrow (2p)^5 4p$ at .. 3390 nm.

There is a particular problem with the removal of atoms from the $(2p)^5 3p$ levels: atoms in these levels decay quickly to the $(2p)^5 3s$ levels, but those levels are metastable. As their populations build up, atoms will be fed back up to the $(2p)^5 3p$ levels via the phenomenon of 'resonance radiation'. This would mean a build-up of atoms in the lower laser level and the destruction of the population inversion. To maintain the population inversion, the atoms must be removed from the metastable states by collision with the walls of a capillary tube mounted along the axis of laser cavity. However the complications of the process mean that the gas pressure and electron density in the laser can only be varied within rather narrow limits so that the small-signal gain of the He–Ne system can be nowhere near that of the argon ion system.

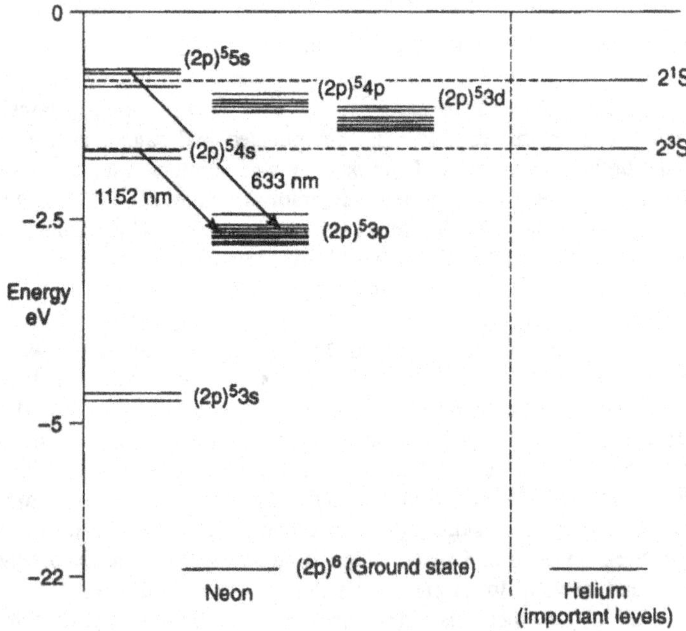

Fig. 10.9 The processes leading to population inversions in neon. The two best known laser transitions are marked, although many others are possible. (The helium levels are drawn with the ground state coinciding with that of neon.)

11

Atoms in external fields

In this Chapter, experiments using external electric and magnetic fields, which enable further investigation of atomic structure, are described. With the development of high vacuum and laser techniques, such experiments have achieved greater and greater importance and many of the most interesting research into atomic physics today is carried out using one or a combination of the techniques discussed below.

11.1 THE STERN–GERLACH EXPERIMENT

One of the earliest and perhaps the most important such experiment is the experiment performed by Stern and Gerlach in 1921 to demonstrate directly the existence of electron spin. Atoms from an oven are formed into a beam which

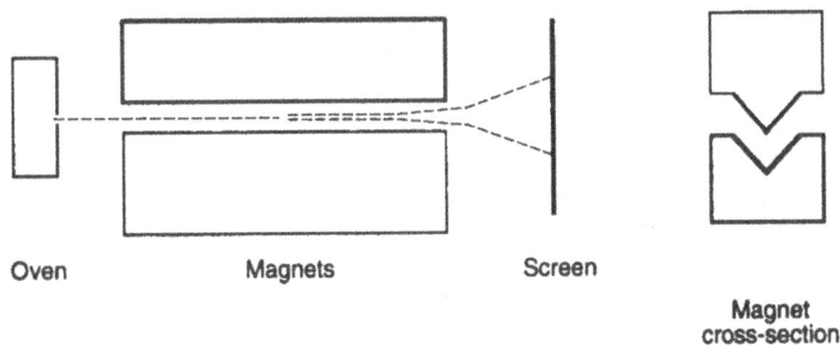

Fig. 11.1 Apparatus for observing the Stern–Gerlach effect. A beam of atoms from a heated oven travels through a highly inhomogeneous magnetic field and hits a screen.

travels in a vacuum chamber between the poles of a magnet. These poles have been specially shaped to give a sharply changing magnetic field in the z-direction (Fig. 11.1). Classically, this means that there will be a force on the atom in the z-direction given by

$$F_z = \boldsymbol{M} \,.\, \mathrm{d}\mathbf{B}/\mathrm{d}z \tag{11.1}$$

where $\boldsymbol{M}$ is the magnetic dipole moment of the atom and $\mathrm{d}\mathbf{B}/\mathrm{d}z$ is the rate of change of the $\boldsymbol{B}$ field in the z-direction.

If the oven contains sodium heated to 1000 K, simple kinetic theory shows that the beam of sodium atoms will, on average, have a speed of some 200 m/s. The vast majority of the atoms will be in the ground state, which is a $^2S_{1/2}$ state.

This means that there are two possible values of the spin state of the atom depending whether the electron is in a spin-up or a spin-down state. The $\boldsymbol{B}$ field provides a force on the atoms which, from (11.1), is either

$$+\tfrac{1}{2}(geh/4\pi m)\ \mathrm{d}B/\mathrm{d}z \quad \text{or} \quad -\tfrac{1}{2}(geh/4\pi m)\ \mathrm{d}B/\mathrm{d}z \quad \mathrm{N}$$

where g, the electron 'g factor' was defined in Section 5.3. Thus the beam is split into two, as shown in Fig. 11.1.

This experiment gives direct evidence for the existence of electron spin and a quantitative measurement of its magnitude. It also demonstrates directly that the energy is quantized and this quantization is defined by the external $\boldsymbol{B}$ field. In other words, the existence of the $\boldsymbol{B}$ field specifies the z-direction for the spin angular momentum operator s_z. A similar result would be obtained for any atom whose ground state is a doublet. Conversely, for atoms such as phosphorus, which has a ground state $^4S_{3/2}$, splitting into four beams would result.

The experiment gave a value for the electron g factor, $g = 2.002$.

11.2 ATOMS IN HOMOGENEOUS MAGNETIC FIELDS – THE ZEEMAN EFFECT

Quite early in the history of experimental spectroscopy it was noted by Zeeman and others that a constant magnetic field imposed on an atomic discharge tube produced a splitting of spectral lines (Fig. 11.2). Once again, this effect is due to the extra energy possessed by the atomic electrons due to the external $\boldsymbol{B}$ field, and obviously energy due to both orbital and spin angular momentum of the atom must be taken into account.

However, by choosing a particular atomic gas discharge, simplification is possible. Consider the transition at 423 nm between an excited state in calcium and its ground state

$$4\mathrm{p}\ ^1P_1 \rightarrow 4\mathrm{s}\ ^1S_0.$$

In this case the total spin of each state is zero, so the only effect of the B field is the effect on the orbital angular momentum L. The energy due to the interaction between the magnetic field B_z and the (orbital) magnetic moment of the atom is given by

$$W = - L_z \, (eh/4\pi m) \, B_z \quad \text{J}. \tag{11.2}$$

For the ground state, L_z is zero. For the excited state L_z may have values +1, 0 or -1. Thus the line is split into three components (Fig. 11.3). In this case, since W is not spatially dependent, time-independent perturbation theory (Appendix C) gives a particularly simple result

$$\Delta E = - L_z \, (eh/4\pi m) \, B_z \int u_n{}^* \, u_n \mathrm{d}v \quad \text{J}.$$

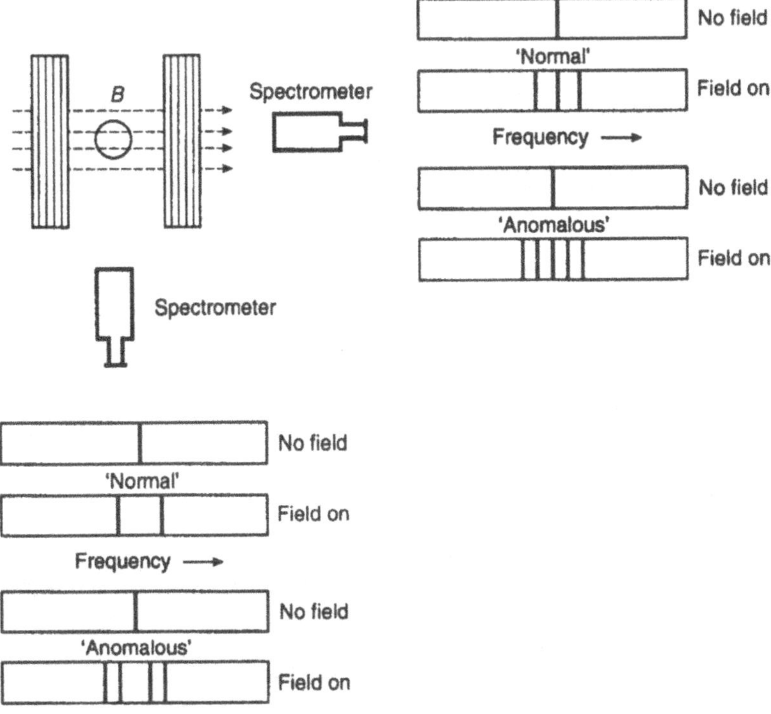

Fig. 11.2 The Zeeman effect. Note that the direction of viewing changes the effect seen.

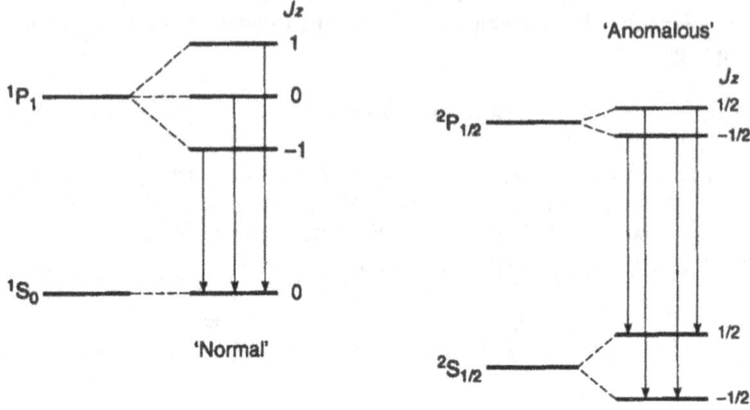

Fig. 11.3 Energy level diagram to illustrate 'normal' and 'anomalous' Zeeman splitting of energy levels. Note that the undeviated $^1P_1 - {}^1S_0$ transition would only be observed in the the direction of the magnetic field as shown in Fig. 11.2.

The integration gives

$$\Delta E = - L_z\,(e\hbar/4\pi m)\,B_z \quad \mathrm{J}. \tag{11.3}$$

It should be noted that the transition equivalent to the unshifted line can only be seen if the discharge is viewed along the direction of the $\boldsymbol{B}$ field. This is purely due to the shape of the radiation field emitted by dipole radiation. For a similar reason the states of polarization of the components viewed along and perpendicular to the $\boldsymbol{B}$ field are different.

This type of Zeeman effect with splitting into three lines is known as the normal Zeeman effect and only occurs for singlet–singlet transitions. It is a small effect. Choosing a magnetic field of 1 tesla and substituting in (11.3) gives a line splitting of the order of 10^{-4} eV.

The 'anomalous' Zeeman effect occurs much more commonly, when other types of transition are observed. The case of a $^2P_{1/2} - {}^2S_{1/2}$ transition, producing four lines, is also shown in Fig. 11.3. Ironically, early experimenters assumed the 'normal' form to be the most common as their instruments did not have a great

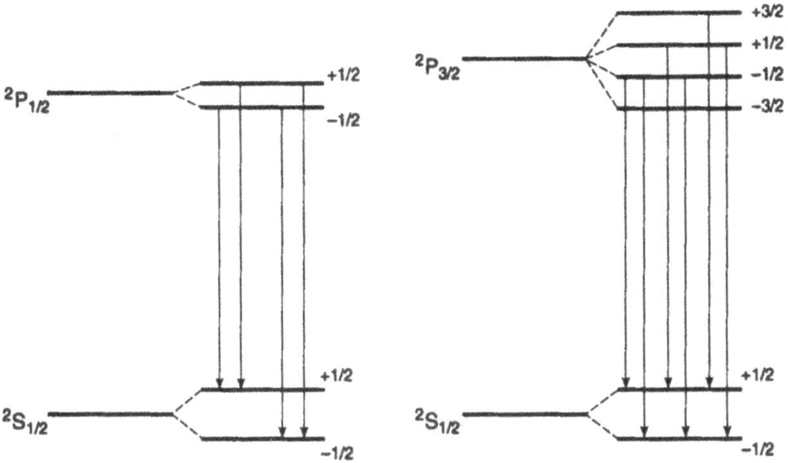

Fig. 11.4. Zeeman effect splitting of the sodium D lines Note the difference in the splitting of the upper and lower states which produces four components for the $^2P_{1/2}$ - $^2S_{1/2}$ transition and six components for the $^2P_{3/2}$ - $^2S_{3/2}$ transition.

enough resolution for them to see the multiple splitting characteristic of 'anomalous' transitions. (Also, normal splitting could be explained by semiclassical models.)

In order to calculate the anomalous Zeeman splitting, the effect of the addition of the electron's orbital and spin angular momenta must again be considered. By extension from (11.2), the extra energy term is now given by

$$W' = -(eh/4\pi m)(L_z + gS_z) B_z \quad \text{J} \qquad (11.4)$$

where g is the electron g factor.

However, for quantum angular momentum operators, when the total angular momentum operator **J** is well defined, L_z and S_z are not well defined (see Appendix B). But the z-component of **J**, J_z is well defined, and it is possible to rewrite (11.4) as

$$W' = -(eh/4\pi m) g_J \, J_z B_z$$

or

$$W' = -(eh/4\pi m) g_J \; \mathbf{J} . \mathbf{B}. \qquad (11.5)$$

The coefficient g_J is known as the Landé g factor. Manipulation along the lines shown in Appendix B gives a value for g_J for any particular transition.

For states when $S = 0$ (for example, doublet states), (11.5) reverts to the form of (11.2). For other states, 'anomalous' splitting occurs as shown in Fig. 11.4.

11.3 STRONG MAGNETIC FIELDS – THE PASCHEN–BACK EFFECT

In an experiment where the anomalous Zeeman effect is observed, increasing the magnetic field increases the splitting linearly, in line with (11.5). However, for large enough fields, the lines produced by the anomalous Zeeman effect coalesce to produce a pattern typical of the normal Zeeman effect. This is because when the magnetic splitting becomes greater than the fine structure splitting, L and S become uncoupled and L_z and S_z in (11.4) become well defined (compare the discussion of coupling in Section 9.7). In this case, the selection rule

$$\Delta S_z = 0$$

is obeyed and since the second term in the equation is almost exactly twice the first term, a normal Zeeman pattern results. Obviously the fine structure terms still produce a small splitting in the new 'normal' components. This effect is known as the Paschen–Back effect and is another example of the vital importance of the relative sizes of the various energy perturbations in the atomic system to the resulting spectral pattern.

11.4 ATOMS IN STATIC ELECTRIC FIELDS – THE STARK EFFECT

It might be assumed that subjecting atoms to constant electric fields would produce an effect similar to the Zeeman effect. It does not, for an interesting reason.

If a constant electric field E_z is imposed on an atom the effect on an atomic electron in an eigenstate u_n can be calculated using time-independent perturbation theory (assuming that the energy due to the field is small compared to that due to the atomic nucleus). Classical e.m. theory gives an expression for the energy due to the field :

$$H' = ez\, E_z$$

where $-ez$ is the dipole moment induced by the external field.

From Appendix C, the change to the eigenvalue E_n is given by

$$E' = \int u_n^* \, H' \, u_n \, \mathrm{d}v. \qquad (11.6)$$

But H' is an odd function and changes sign under inversion. On the other hand, atomic eigenfunctions are even functions and do not change sign. Therefore E' must always be zero and there is no linear Stark effect for atoms due to a static electric field, except in one special case which will be noted shortly. Some molecules do exhibit a linear Stark effect, and this can be seen simply in terms of the centre of charge of the molecular electron cloud not coinciding with that of the positive nuclei, thus producing a static electric dipole moment which will interact with the static electric field.

On the other hand, (11.6) only expresses the first order correction due to H'. Taking the perturbation analysis of Appendix C to higher orders, produces a second order correction to E_n. This will be a function of H'^2, which is even. Thus there is a 'quadratic' effect, depending on E_z^2. However the effect is very small.

There is one special case where an atomic linear Stark effect does occur. This is when degenerate eigenstates exist. In this case the simple time-independent analysis cannot be used, as can be seen by examining equations C.5 and C.6. A more sophisticated form of analysis is required, as shown in Appendix C.

If the n = 2 states of hydrogen are considered, the degeneracy of the four states means that the perturbation energy is not zero and so a linear Stark effect occurs. In physical terms, the perturbation produces states which are a combination of s and p orbitals, which can be seen to have a non-centrally symmetric electron cloud. However the effect is small. For a very strong electric field of 10^6 V/m, a splitting of the order of the order of 10^{-4} eV occurs.

11.5 INTENSE ELECTROMAGNETIC FIELDS – RABI OSCILLATION

The perturbation analysis of the interaction of an atom with an electromagnetic field given in Chapter 6 relied on the fact that the probability of transition of the initial atomic state to another state was small. This implied that the initial state population of atoms was not greatly changed by the interaction with the field.

The invention of the laser made possible experiments where neither of these applied. It became possible to study the interaction of atoms with an extremely intense field, at a frequency ν very close to that of an atomic transition ν_0, which had a frequency divergence smaller than the natural linewidth of the transition $\Delta\nu_0$. In such a case a resonant interaction will cause large changes in the atomic population. Because this interaction is between the atom and the electric field vector of the wave, it is sometimes known as the 'dynamic Stark effect'.

The theoretical analysis of such a system is, in fact, rather simpler than the perturbation approach, and had been worked out in the 1930s by Rabi for the case of particles with intrinsic spins interacting with an oscillating magnetic field.

Consider the picture given in Appendix D, where the extra energy term H' is now large compared with H. For two states defined by the subscripts i and f, the time-dependent probabilities of finding atoms in these states is defined by the coefficients $a_i(t)$ and $a_f(t)$ (as in the previous analysis, the atom starts in state i, so $a_i(0) = 1$ and $a_f(0) = 0$). If transitions may only take place between these two states, it is possible to write down directly simple coupled equations relating a_i and a_f

$$i(h/2\pi)\, da_i/dt = D_{12} a_f \exp(-i\omega_T t) \cos(2\pi\nu t)$$

$$i(h/2\pi)\, da_f/dt = D_{12} a_i \exp(+i\omega_T t) \cos(2\pi\nu t) \quad (11.7)$$

where $\omega_T = 2\pi\nu_0 = 2\pi(E_f - E_i)/h$, and D_{12} is the dipole moment matrix element discussed in Appendix D.

The solutions for these equations are easily worked out. To simplify even more, taking the laser frequency ν to be equal to ν_0, a simple oscillating solution is found for a_f. Thus the variation with time of the probability of finding an atom

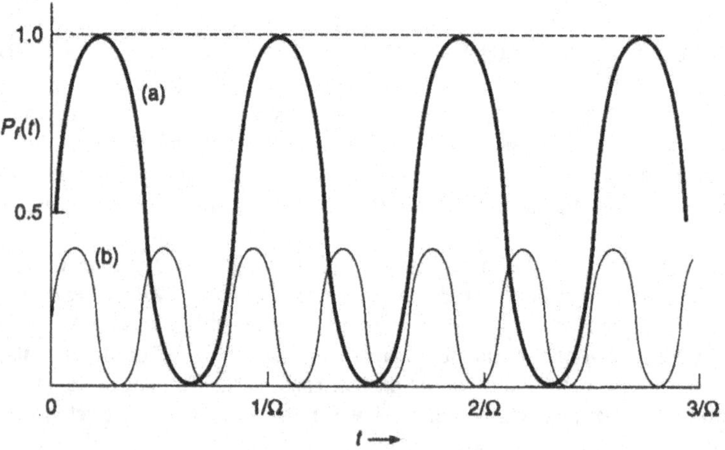

Fig. 11.5 Rabi oscillations for an atom in an intense field. The resonant case where $\nu = \nu_0$ (a) and a just off resonant case (b) are shown.

in state f, $P_f(t)$ may be written down

$$P_f(t) = |a_f|^2 = \sin(2\pi^2 |D_{12}| /h)t. \quad (11.8)$$

This means that the atom oscillates sinusoidally between the initial and final states with a frequency Ω defined only by the strength of the field–atom interaction.

If the field is detuned from resonance, as would be expected, the maximum value of $P_f(t)$ decreases and the period of oscillation increases. In this case, the expression for Ω is given by

$$\Omega = {}^1/_2\,[(\nu - \nu_0)^2 + 4\pi^2|D_{12}|^2/h^2)]^{1/2}\ \text{Hz}. \quad (11.9)$$

Rabi oscillation is shown in Fig. 11.5 for the resonant and just off resonant cases. It can be seen that in either case, the value of Ω is dependent on the amplitude of the radiation field. The perturbation solution is regained if D_{12} is so small that the second term in the bracket in (11.9) is negligible compared to the first.

These strong interactions between electromagnetic fields and atomic systems can be used to produce many interesting effects. Pulse propagation through the medium can result in phenomena where pulses propagate through a medium without changing shape, an effect known as self-induced transparency.

11.6 ATOMIC BEAM EXPERIMENTS

The vast majority of the primary experimental investigations of atomic structure were carried out using gas discharge tubes. As explained in Chapter 4, these have the advantage of simplicity and the fact that the mean free time of an atom can be made very large compared with the times of interest in atomic transitions. However, the random motion of atoms produces a major disadvantage since Doppler effects are always present to mask investigation of fine structure. Also, the gas pressure must be kept relatively low in order to minimize collision broadening and this puts a limit on the intensity of radiation emitted. The Stern–Gerlach experiment is an early example of another technique which gets round these limitations, the use of an atomic beam.

A typical system for producing such a beam is shown in Fig. 11.6. A small oven containing a sample of sodium is heated to a high enough temperature to give a reasonable vapour pressure of sodium gas, producing a beam of sodium atoms which issues from the aperture in the oven. The beam density increases linearly with temperature until an optimum temperature is reached. This is defined by the constraint that the mean free path of the atoms must be greater than the dimensions of the aperture in the oven from which the beam emerges.

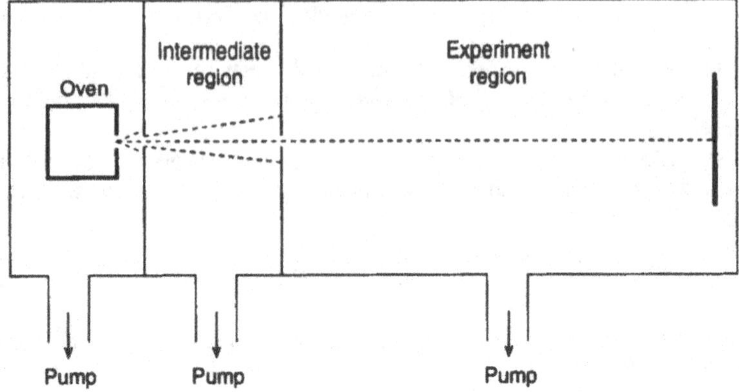

Fig 11.6 Typical apparatus for the production of atomic beams. The beam of atoms from the oven is collimated in the intermediate region.

Above this temperature, the beam density is constant. In the case of sodium, a slit of width 0.1 mm and height 10 mm requires an oven temperature of 670 K. Such a set-up can be used to produce atomic beams for a very large number of elements, with different experimental arrangements depending on whether the element is normally a solid, liquid or gas. With high enough slit temperatures, molecules such as Bi_2 and H_2 can be dissociated to give beams of their constituent atoms. Ion beams may also be produced.

The beam is collimated so that it is directed along the axis of the apparatus. Collisions are minimized and only take place in the unlikely event of a faster atom catching up a slower one. However, the use of an oven aperture which provides supersonic flow provides a beam in which the spread of velocities, and hence collisions is very much smaller (Fig. 11.7). It is possible to use 'crossed beam' techniques, where the interaction of two beams enables detection of the results of interactions between atoms of well-defined speed and position.

A major use for atomic beams has been the study of reactions between electrons, ions, atoms and molecules. There are two important types of study which produce information on reactions familiar from the discussion of lasers in Chapter 10. Electron impact ionization, which may be represented by the equation

$$e + \mathrm{A} \rightarrow \mathrm{A}^+ + 2e$$

occurs in the production of a population inversion in the argon ion laser. Charge

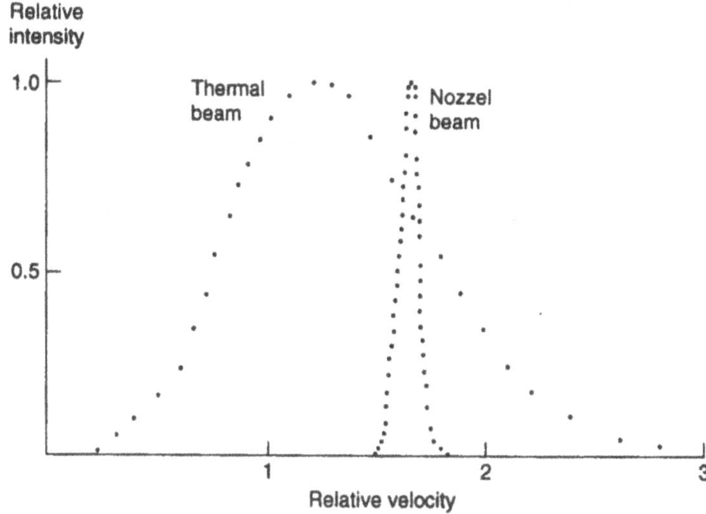

Fig. 11.7 Velocity distributions normalized to the same peak intensity for a thermal beam and a supersonic nozzle beam operating at Mach 5 (Reproduced with permission from Pendlebury, J.M. and Smith, K.F., *Contemp. Phys.*, *28*(1), page 3.(1987).

transfer reactions, which may be represented by

$$A^+ + B \rightarrow A + B^+$$

also occur in laser systems.

More importantly, the development of the original Stern–Gerlach apparatus can give precise values for atomic parameters by studying the frequencies associated with transitions between levels. The atomic beam magnetic resonance apparatus first conceived by Rabi enables transitions between levels in the hyperfine structure of the atomic system which could never be resolved by optical spectroscopy.

The principle is straightforward and a typical system is shown in Fig. 11.8. A collimated beam of atoms passes through a strongly inhomogeneous magnetic field similar to that in the Stern–Gerlach experiment. It then passes through a constant magnetic field which is perturbed by a radio-frequency oscillating magnetic field. Finally the beam traverses a second inhomogeneous magnetic field identical to the first, but inverted. Atoms with a particular magnetic moment will only reach the detector if the oscillating magnetic field has caused a transition which reverses the direction of this moment. Obviously the frequency at which this occurs gives the energy difference between the levels

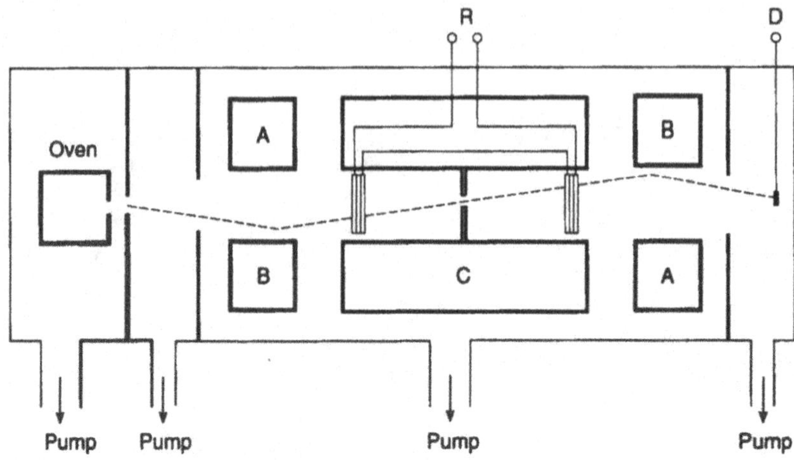

Fig. 11.8 Basic atomic beam magnetic resonance apparatus. Atoms follow the path shown when transitions between states produced by the oscillating RF magnetic field result in opposite deflection by the inhomogeneous magnetic fields between A and B. The homogeneous magnetic field provided by magnet C enables the effect of the magnetic field on the transitions to be observed.

involved. If the atomic nucleus has a magnetic dipole moment m_I with states which can be described by a quantum number I, by analogy with the discussion of fine structure, it can be seen that each hyperfine state can be described by a quantum number F which can take values between $I + J$ and $|I - J|$. Thus there are $(2F + 1)$ hyperfine levels and atomic beam magnetic resonance can measure their energy splitting. This coupling only holds good when the external oscillating magnetic field is small. As the field is increased, the coupling between the electron and nucleus is broken. Studies of the change in resonance frequency as the external field is increased enable values of the nuclear g factor, g_I and other fundamental quantities to be measured.

A development of the system working on the splitting between the two hyperfine levels in the ground state of caesium ($\sim 9 \times 10^9$ Hz) can be used as part of a feedback oscillator with a reproducible precision of better than 1 part in 10^{12}. This is used as a fundamental standard of frequency and is precise enough to be used in experimental tests of general relativity.

Finally, the combination of atomic beams and lasers enables spectroscopy to be carried out free from the restrictions on precision imposed by the Doppler spreading of absorption or emission lines. The combination of tunable and essentially monochromatic laser radiation with atomic beams enables experiments to be carried out to a precision undreamed of in earlier years.

11.7 THE LAMB SHIFT IN HYDROGEN

The first section of this chapter described an experiment earth-shattering in its implications. It is fitting to end with one which had almost as far reaching results, which are still the subject of research.

In the discussion of the fine structure of hydrogen (Section 5.4) it was mentioned that when spin–orbit and relativistic effects are included, the $2^2P_{1/2}$ and $2^2S_{1/2}$ states are energy degenerate. However, when zero point energy of the quantized electromagnetic field is taken into account, the S state is raised and there is a splitting of approximately 4×10^{-6} eV (corresponding to a frequency of about 1000 MHz). Lamb and Retherford used an atomic beam method to measure this splitting in the apparatus shown in Fig. 11.9.

Hydrogen atoms excited to the $2^2S_{1/2}$ state by electron bombardment pass through an oscillating field in a coaxial line to a detector. The detector works on the basis that excited hydrogen atoms colliding with the surface give up energy. This energy enables electrons to be released from the surface, producing a current which can be measured (a process exactly analogous to photoelectric detection).

When the frequency of the oscillating field is exactly that of the $2^2P_{1/2} - 2^2S_{1/2}$ transition, a proportion of the $2^2S_{1/2}$ atoms is converted to $2^2P_{1/2}$. However, unlike the $2^2S_{1/2}$ state which is metastable, the $2^2P_{1/2}$ state has a very short lifetime for

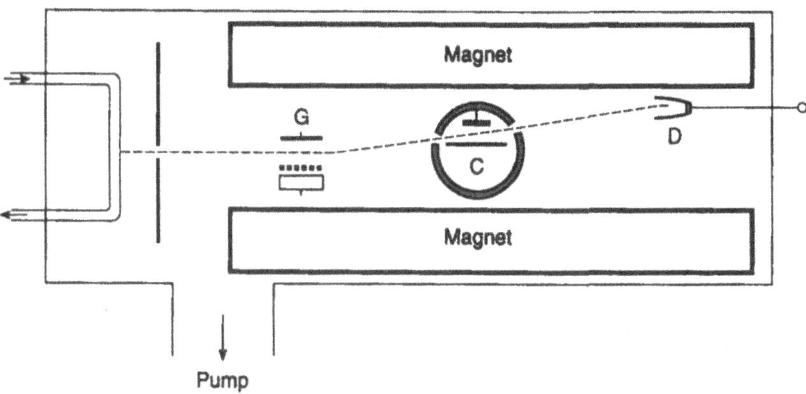

Fig. 11.9 Apparatus for measuring the Lamb shift in hydrogen. Hydrogen flows through the heated tube and a collimated beam of atoms is produced. A small fraction of these are excited by electron bombardment at G. The beam then passes through the R.F. field in the coaxial cable C and reaches the detector D where only excited atoms are registered. The static magnetic field is used to investigate the field dependence of the transitions.

decay to the ground state. Thus atoms in this state decay to the ground state before they reach the detector and ground state atoms will not release electrons. Resonance is therefore demonstrated by a fall in the emission current in the detector. The shift between the two levels (now known as the Lamb shift) has been measured precisely as 1058 MHz in perfect agreement with the prediction of quantum electrodynamics. More recent work has given a measurement of the Lamb shift to 1 part in 50 000. This gives a searching test of QED.

12

Single atom experiments

In the 1970s and 1980s technological developments, notably the development of lasers but also advances in ultra-high vacuum techniques and electronics, produced a revolution in atomic physics comparable with that caused by the development of spectroscopy in the 1870s. Today, experiments can be carried out which would have been considered impossible thirty years ago. Single atoms can be observed and experimented on and many new and strange phenomena investigated. This chapter gives a sketch of some of these developments.

12.1 EXPERIMENTS ON SINGLE ELECTRONS AND IONS

The first experiments to be discussed would have been quite comprehensible to J.J. Thomson. Dehmelt and his co-workers devised a way of 'trapping' electrons

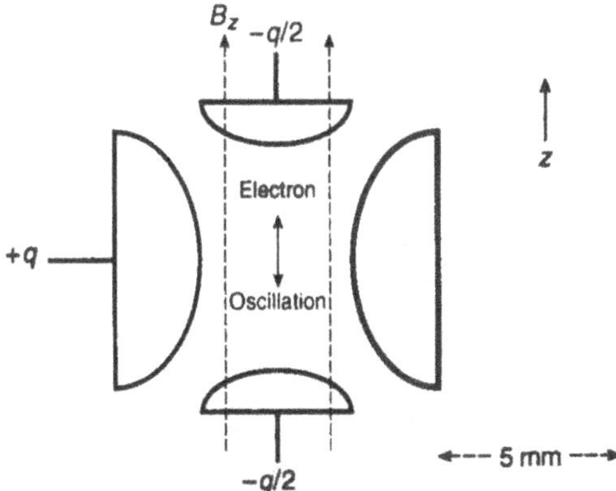

Fig. 12.1 Penning trap. The trap has cylindrical symmetry about the z-axis. The ring-shaped electrode carries a positive charge. An electron will oscillate along the z-axis.

using electric and magnetic fields. The device, known as a 'Penning trap' is shown in Fig 12.1. A weak electric field, together with a strong magnetic field produce a 'saddle point' in the electric potential energy at the centre of the trap. The shape of the electrodes means that the potential energy of the electron is given by

$$V = A\,(x^2 + y^2 - 2z^2) \tag{12.1}$$

where A is a positive constant. Thus for electrons on the axis of the trap, there is a harmonic restoring force $F_z = -4eAz$. and the electron will oscillate in the z-direction with frequency

$$\nu_z = \surd(eA/\pi m)\ \text{Hz}. \tag{12.2}$$

At the same time the electron moves in a circular orbit around the lines of force of the B field. This 'cyclotron motion' has a frequency

$$\nu_c = eB/\,2\pi mc\ \text{Hz}. \tag{12.3}$$

The electric field slightly perturbs this motion so that the actual frequency is fractionally lower.

If the trap is in ultra high vacuum (residual gas density of order 1 atom/cc), electrons can be injected into it and will remain in orbit for the order of minutes. They may be detected via the electromagnetic radiation they emit, which is in

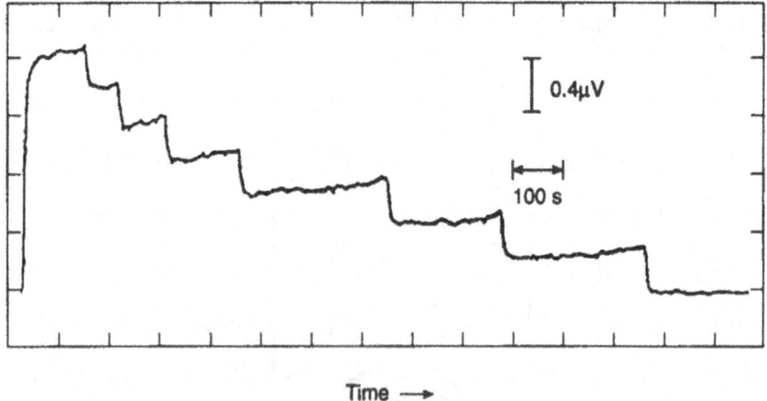

Fig. 12.2 R.F signals produced by electrons in a Penning trap. Seven electrons are initially injected into the trap and diffuse out, one at a time, over a period of the order of minutes. (Reproduced with permission from Wineland, Ekstrom and Dehmelt, *Phys. Rev. Letts.* 31, 1279 (1973).)

the MHz region (Fig. 12.2). In fact, the orbit tends to decay so that a 'cooling' technique, similar to those discussed later, is required to keep a single electron in oscillation for a long period.

A weak inhomogeneous magnetic field can be added which will slightly perturb the electron motion due to its interaction with the electron's spin magnetic moment. This perturbation will be different depending on whether the electron is in a 'spin-up' or a 'spin-down' state, so the measurement of the difference frequency enables a very precise measurement of the electron g factor (Section 5.3) In fact the g factor can be measured to a precision of 4 parts in 10^{12} as

$$g = 2.002\,319\,304\,376.$$

The precision of this experimental value enables a very stringent test of quantum electrodynamics. In passing, although the theory of fundamental particles is not part of the present discussion, it is worth mentioning that this result, together with QED, suggests a value for the electron radius of 10^{-24}m! It can be seen that experiments capable of this precision are able seriously to test fundamental theoretical pictures of matter.

Obviously, the experimental techniques described above can be applied equally to ions. Most interestingly, a single ion may be trapped and then excited by a laser beam from outside the system. Spontaneous emission from this ion may then be observed – seeing a single atom directly!

12.2 TRAPPING NEUTRAL ATOMS

The Penning trap obviously only works on charged particles. However a different technique enables individual neutral atoms to be trapped and experimented on. This technique is known as 'laser cooling'.

Returning to Einstein's model discussed in Section 6.3, the interaction between an atom and a beam of radiation travelling in the x-direction involves not only an exchange of energy but an exchange of momentum. In essence, the absorption of one quantum of radiation by the atom must be accompanied by the absorption of an equivalent momentum and the changing of the atom's velocity in the direction of the radiation by an amount

$$\upsilon_x = hc/\lambda M \qquad \text{M/s} \tag{12.4}$$

where M is the mass of the atom and λ the wavelength of the field. The atom will lose the excess energy by spontaneous emission. The probability of this emission in the positive and negative x-directions will be equal, so that on average over a number of interactions between the atom and the field there is no net change in momentum in this process. At high laser intensities, the interaction rate will be approximately equal to $1/2\tau$, where τ is the natural lifetime of the

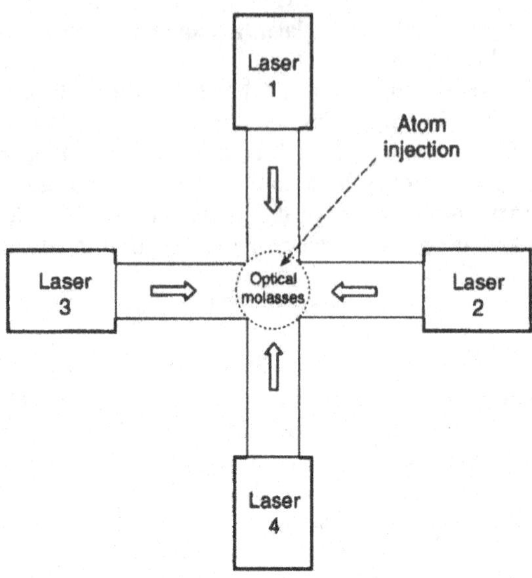

Fig. 12.3 Laser set-up to produce 'optical molasses' for the cooling and trapping of neutral atoms. A third pair of lasers is perpendicular to the plane of the paper.

excited state.

The final trick required to trap an atom comes from the Doppler shifting of the radiation frequencies discussed in Section 5.5. Remembering that the resonant frequency of a moving atom is effectively shifted by its relative motion, if a gas of atoms interacts with a laser field which has a frequency just below the unshifted atomic resonant frequency, only those atoms which are moving in the opposite direction to the field will interact. Thus the momentum transfer between the atom and the field will tend to slow down the atom.

If a three-dimensional array of lasers is set up, tuned to the appropriate frequency, atoms injected into the centre of the combined field move more and more sluggishly (Fig 12.3). The combined laser field acts as a very viscous medium (christened 'optical molasses') and the atoms drift very slowly through the volume. Although the 'optical molasses' does not constitute an actual trap, it captures and slows atoms. The atomic speeds are quoted in terms of the equivalent temperature of the gas of atoms. Effective temperatures of the order of 10 μK can be achieved, and the atoms remain in the volume for some seconds before diffusing out.

True trapping can be achieved if atoms which have magnetic dipole moments are used (for example, sodium atoms). For such atoms, the minimum point in a quadrupole magnetic field can act as a trap, in a similar way to an electric field in a Penning trap. However, such a trap is rather shallow, so that atoms at normal temperatures would have enough kinetic energy to escape from it. If the atoms are first cooled, using 'laser molasses', they will not have enough energy to escape.

There are numerous experimental refinements to the techniques described above. The upshot is that the trapping and observation of a few atoms or even a single atom over periods of seconds, or even hours, is quite feasible.

Once trapped, what sort of experiments can be carried out on the atoms? One particular experiment will be described because of its relevance to the atomic physics discussed previously.

12.3 THE QUANTUM 'WATCHED POT'

This experiment is named from the proverb 'a watched pot never boils'. In more elevated company it is known as the quantum Zeno effect after one of the paradoxes proposed by the ancient Greek philosopher. It is a good demonstration of the sort of experiment now possible.

The apparent paradox arises from the discussion of atomic transitions in Section 6.4. As before, consider an atom which may be in one of two optically connected states, 1 and 2 ($E_2 > E_1$). The state function for the atom consists of two parts, one referring to eigenstate 1 and the other to eigenstate 2 as in (6.12)

$$\Psi = c_1(t)\, u_1 \exp[-i(2\pi E_1 t/h)] + c_2(t)\, u_2 \exp[-i(2\pi E_2 t/h)]. \qquad (12.5)$$

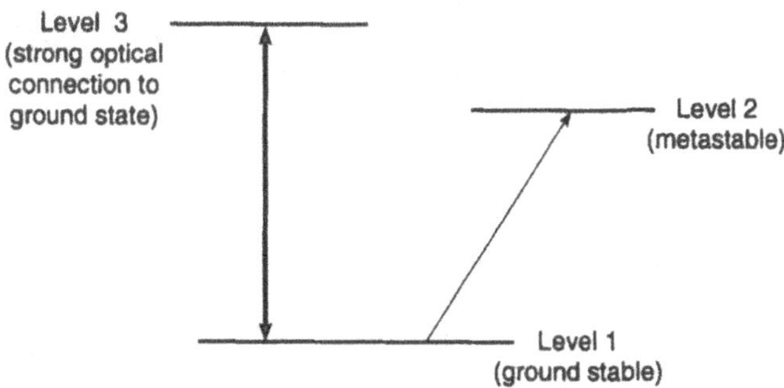

Fig. 12.4 Simplified energy level diagram for a barium ion quantum 'watched pot' experiment.

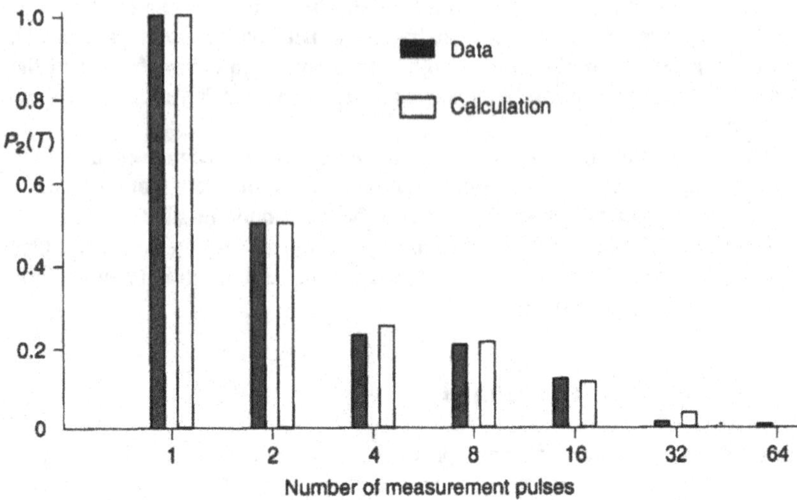

Fig. 12.5 The 'watched pot'. Graph of the probability of finding the ion in state 2 after a time T as a function of the number of observations during T. (Redrawn from Itano, Heinzen, Bollinger and Winelund, *Phys. Rev. A*, 41, 2295, (1990)).

Assume that at time $t = 0$ the atom is in the lower state, so that $c_1(0) = 1$ and $c_2(0) = 0$. A simple picture of the transition process would have $c_1(t)$ decreasing with time and $c_2(t)$ increasing, so that at some long time T later, it is possible to say with confidence that $c_2(T) = 1$ and $c_1(T) = 0$. However, if it were possible to inspect the atom at some time t' ($< T$) and if the atom were found still to be in state 1, it would be obvious that $c_1(t')$ had been 'reset' to 1 and $c_2(t')$ to 0. In other words, the decay process had been 'reset to zero'. The conclusion is that if n measurements are made, spaced in time by T/n, the probability that the atom remains in state 1 goes towards unity as n goes towards infinity. Hence a continuously observed atom will never decay!

The experiment was successfully carried out in 1990 by Itano and co-workers using Be^+ ions in a Penning trap. About 5000 ions were trapped and the pressure in the trap was about 10^{-8} Pa, giving a storage time of several hours for the ions. In essence three levels are used (Fig 12.4). Level 1 is the ground state. Level 2 is a metastable state, so that the spontaneous decay from level 2 to level 1 is negligible. However, if an intense radio frequency field with frequency $\nu_{12} = (E_2 - E_1)/h$ is applied to the ion, it can be switched between levels 1 and 2 by the phenomenon of Rabi resonance oscillation (see Section 11.5). If the ion is in level 1 at time $t = 0$, it can be shown that the probabilities P_1 and P_2 of the ion being in states 1 and 2 are given by

$$P_1(t) = \cos^2(\Omega t/2) \; ; \; P_2(t) = \sin^2(\Omega t/2)$$

where Ω is the Rabi frequency and is proportional to the amplitude of the field. If a measurement of the state of the ion is made after a short time T such that $\Omega T < 1$, then $P_1(t) = 1$ and $P_2(t) << 1$.

How is the state of the ion detected? This is where level 3 comes in. This level is connected to level 1 by a strongly allowed transition, and can only decay to level 1. If a short optical pulse of frequency ν_{13} hits the ion, it acts as a probe for the state of the ion. If the ion is in level 1, the probe pulse causes the ion to cycle between levels 1 and 3 emitting radiation at frequency ν_{13}. If, on the other hand, the ion is in level 2, no interaction occurs and no light is emitted. So, the pulse acts as a non-destructive test of the state of the ion. If the atom is in level 1, the probe causes the emission of light at frequency ν_{13} and the atom returns to level 1 when the probe is switched off. If, on the other hand, the ion is in level 2, it remains there undisturbed. So, if the probe pulse is switched on at time t, probing 'collapses' (12.5) so that $c_1(t)$ and c_2 (t) are 'set' to 1 or 0, even though the probe itself has not affected the state of the ion.

Experimentally, a signal at ν_{13} is first applied to set the ion in level 1. The RF excitation is switched on for a time T. During this time, n probe pulses of length τ ($\tau << T$) are applied. At the end of time T, the probe pulse is turned on again and the number of quanta of light emitted gives a measure of the number of ions remaining in level 1 and hence the probability that transition to level 2 has occurred. Figure 12.5 shows how the transition probability depends on the number of probe pulses.

The above experiment is a good example of the sort 'single atom spectroscopy' which can now be carried out.

12.4 CAVITY QUANTUM ELECTRODYNAMICS

Experiments of the type discussed in the previous section demonstrate that concepts such as 'decay time', which might have been thought of as intrinsic properties of the atom are, in fact, defined by the interaction of the atom with the system in which it finds itself. (This might also have been gathered from the discussion of spontaneous emission, in Section 6.8.) Single atom experiments enable a whole new field of phenomena to be investigated. Only a few examples of this fascinating subject can be mentioned here.

Returning to Section 6.8, it was necessary to associate a 'zero point energy' $^1/_2 h\nu$ J with each mode of the radiation field. But the number of modes in a field is defined by the boundary conditions on the field. As long ago as 1948, Casimir pointed out that the number of field modes between two plane parallel plates could be easily calculated and was different from those in the space outside the plates. Therefore the 'zero point' field energy between the plates was less than that outside and so a 'zero point' force should exist between the plates. This effect is tiny, and has not yet been convincingly observed. However, in experiments where a single atom can be placed between the plates, the finite number of field modes with which it can interact will produce a change in its natural lifetime and in the radiation pattern emitted. Experiments have been carried out demonstrating that (for example) the natural lifetime of a sodium atom can be reduced when it is allowed only to interact with a single mode of the radiation field in a cavity. Conversely, if an atom is isolated in a waveguide which has a cut-off frequency greater than the atom's transition frequency, no modes of the radiation field exist for the atom to interact with and the atom's lifetime is increased dramatically. This effect has also been observed.

It is possible also to measure very precisely shifts in energy levels due to the interactions between atoms and quantized electromagnetic fields in cavities. These measurements make possible very sensitive tests of the predictions of quantum electrodynamics, but are beyond the scope of this book.

Appendix A

Rutherford's scattering formula

In Section 2.3, Rutherford's analysis of the scattering of particles by a nuclear atom led to the formula for f, the fraction of particles scattered to angles greater than a specific angle Θ

$$f = \pi N t\,(Ze^2/4\pi\varepsilon_0 T)^2 \cot^2(\Theta/2) \tag{A.1}$$

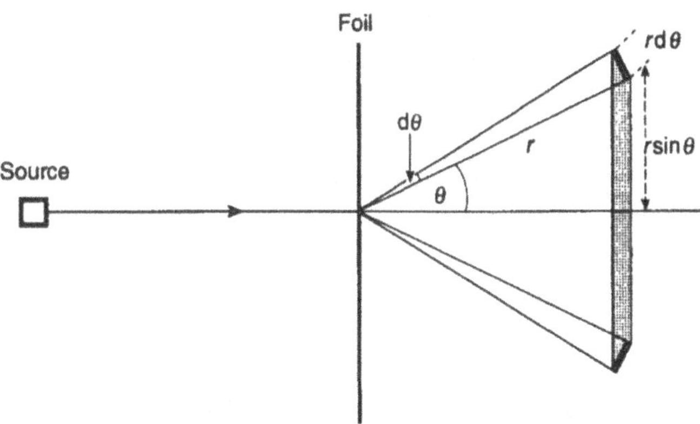

Fig. A.1 Geometrical picture of scattering for the calculation of Rutherford's formula.

where N is the number of atoms per unit volume in a foil of thickness t, Ze is the charge on the atomic nucleus and T the kinetic energy of the particles.

In an experimental situation, a detector of area S will be situated as shown in Fig. A.1 at a distance r from the foil. The detector subtends an angle $d\Theta$ as shown. In three dimensions, the fraction of α particles deflected between Θ and $\Theta + d\Theta$ is given by differentiating (A.1) with respect to Θ

$$df/d\Theta = -\pi N t\,(Ze^2/4\pi\varepsilon_0 T)^2 \cot(\Theta/2)\sin^{-2}(\Theta/2). \qquad \text{(A.2)}$$

As shown in Fig. A.1, these particles will fall symmetrically on an area

$$\begin{aligned} A &= 2\pi\, r \sin\Theta \; r\, d\Theta \\ &= 2\pi\, r^2 \sin\Theta\, d\Theta \\ &= 4\pi\, r^2 \sin(\Theta/2)\cos(\Theta/2)\, d\Theta. \end{aligned} \qquad \text{(A.3)}$$

Therefore, if the number of α particles falling on the foil per unit area is n_i and assuming that the area of the foil irradiated by the beam is small compared with r, (A.2) and (A.3) give the number of particles arriving at the detector per unit time as

$$\begin{aligned} n_d &= (S/A)\, df/d\Theta \\ &= n_i\, SNt\,(Ze^2/8\pi\varepsilon_0\, r\, T)^2 \sin^{-4}(\Theta/2). \end{aligned} \qquad \text{(A.4)}$$

Thus

$$n_d \propto \sin^{-4}(\Theta/2). \qquad \text{(A.5)}$$

This equation is in a form which makes it possible for it to be tested experimentally.

Appendix B

Angular momentum operators

Angular momentum plays such an important role in atomic physics that the major results of the quantum mechanical analysis are gathered together in this appendix. Details of the calculations leading to the results are not given, but may be found in quantum mechanics texts mentioned in Appendix I.

B.1 GENERAL DEFINITION

Classically, the general definition of the angular momentum of a particle about the origin is related to its linear momentum $\boldsymbol{p}$ and to its vector distance from the origin $\boldsymbol{r}$.

$$\boldsymbol{L} = \boldsymbol{r} \times \boldsymbol{p}. \tag{B.1}$$

(Note the symbol $\boldsymbol{L}$ is used here for consistency with normal usage.)

In Cartesian co-ordinates, this gives expressions for the three components of $\boldsymbol{L}$

$$\begin{aligned} L_x &= yp_z - zp_y \\ L_y &= zp_x - xp_z \\ L_z &= xp_y - yp_x \end{aligned} \tag{B.2}$$

and the amplitude of the angular momentum vector $|\boldsymbol{L}|$ is given by

$$|\boldsymbol{L}|^2 = L_x^2 + L_y^2 + L_z^2$$

The conversion to the quantum angular momentum operator $\mathbf{L}$ follows from (B.2) by replacing the variables $\boldsymbol{r}$ and $\boldsymbol{p}$ by the operators $\mathbf{r}$ and $\mathbf{p}$ and using the usual expressions for these operators ($\mathbf{x} \rightarrow x$, $\mathrm{p}_x \rightarrow -\mathrm{i}(\mathrm{h}/2\pi)\partial/\partial x$ etc.). An important consequence follows when the commutators of the various components are calculated. It is found that

$$[\,|\mathbf{L}|^2\,,\ \mathrm{L}_x] = [\,|\mathbf{L}|^2\,,\ \mathrm{L}_y] = [\,|\mathbf{L}|^2\,,\ \mathrm{L}_z] = 0. \tag{B.3}$$

All other commutators (for example $[L_x, L_y]$) do not equal zero. This means physically that it is only possible to measure precisely, at the same time, one component of **L** and the amplitude $|\mathbf{L}|^2$. Thus the *z*-component of **L** may be known precisely, as may its amplitude, but its *x* and *y* components will be imprecise. Since many important physical situations have spherical symmetry, the *z*-direction is taken as the defining direction and the quantum mechanical situation is described in terms of a cone with the values of the Θ and ϕ components unspecified.

This result has important consequences. Because they commute, $|\mathbf{L}|^2$ and L_z must have a common set of eigenfunctions. Therefore their eigenequations may be written as follows

$$L_z X(m,b) = m(h/2\pi) X(m,b) \tag{B.4}$$

$$|\mathbf{L}|^2 X(m,b) = b(h^2/4\pi^2) X(m,b). \tag{B.5}$$

The term $mh/2\pi$ is the eigenvalue of the *z*-component of the angular momentum, and $b(h^2/4\pi^2)$ is the eigenvalue of the square of the amplitude of the angular momentum. Considerations of units shows that *m* and *b* are

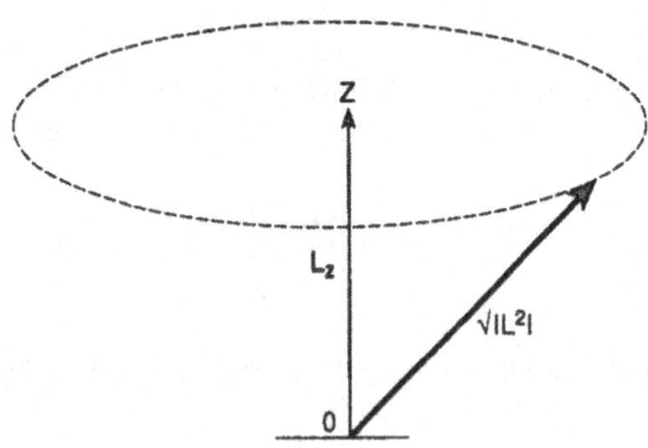

Fig. B.1 Representation of quantized angular momentum vectors. Only the values of the amplitude $\sqrt{|L^2|}$ and the *z*-component L_z can be defined and are shown as solid lines. The θ and ϕ components cannot be specified, so that the amplitude can be thought of as having some random position on the surface of a cone centered on the *z*-axis with apex at the origin.

dimensionless. Since the state X is an eigenstate of both operators, parameter m can be taken as a quantum number which will define possible states of the z-component. Similarly, b will define possible states of the square of the amplitude. The problem is to find possible values of m and b.

The starting point is the fact that for a given value of b (defining the amplitude), there will be a restricted range of values for the z-component and hence for m, bounded by two values m_{max} and m_{min}. It is possible to define two new operators, the raising operator $\mathbf{L}_+$ and the lowering operator $\mathbf{L}_-$, where

$$\mathbf{L}_+ = \mathbf{L}_x + \mathrm{i}\,\mathbf{L}_y \quad \text{and} \quad \mathbf{L}_- = \mathbf{L}_x - \mathrm{i}\,\mathbf{L}_y.$$

Operating with the first of these on the state $X(m,b)$ can be shown to produce a state $X(m+1,b)$, and with the second, a state $X(m-1,b)$. This shows that changes in the z-component of angular momentum are quantized in units of h.

By definition, two equations can be written down

$$\mathbf{L}_+ X(m_{max},b) = \mathbf{L}_- X(m_{min},b) = 0.$$

Using the above expressions, a little mathematical manipulation shows that

$$\mathbf{L}_- \mathbf{L}_+ X(m_{max},b) = (|\mathbf{L}|^2 - \mathbf{L}_z{}^2 - (\mathrm{h}/2\pi)\mathbf{L}_z)\, X(m_{max},b)$$

Using (B.4) and (B.5)

$$\mathbf{L}_- \mathbf{L}_+ X(m_{max},b) = (b - m^2_{max} - m_{max})\,(\mathrm{h}^2/4\pi^2)\, X(m_{max},b)$$

But

$$\mathbf{L}_+ X(m_{max},b) = 0,$$

so that, assuming the state specified by $X(m_{max}, b)$ exists,

$$b = m_{max}(m_{max} + 1)$$

A similar operation on $X(m_{min},b)$ shows that

$$b = m_{min}(m_{min} - 1)$$

and hence that

$$m_{min} = - m_{max}.$$

Finally, since the value of m can only be raised or lowered by integral amounts, the extreme values of m can only differ by an integer so that

$$m_{min} - m_{max} = 2j \tag{B.6}$$

where the only possible values of j are

$$j = 0, 1/2, 1, 3/2, 2, 5/2 \text{ etc.} \tag{B.7}$$

It immediately follows that the value of b is given by

$$b = j\,(j+1).$$

To sum up, it can be seen that if the angular momentum operator is defined via the commutation relationships (B.2), the eigenvalue of its amplitude is defined by the quantum number j, which may only take integral or half-integral values. For a given amplitude, there are a number of possible values of the z-component of the angular momentum, defined by the quantum number m, where m may have any integral value between j and $-j$. Therefore, for completeness, the eigenequations (B.3) and (B.4) may be rewritten

$$L_z\, X(m, j) = m(h/2\pi)\, X(m, j) \tag{B.8}$$

$$|L|^2\, X(m, j) = j\,(j+1)\,(h^2/4\pi^2)\, X(m, j) \tag{B.9}$$

where possible values of j are defined by (B.6) and for a particular value of j there may be values of m

$$m = j,\ (j - 1),\ (j - 2)....0....(-j + 2),\ (-j + 1),\ -j.$$

B.2 ORBITAL ANGULAR MOMENTUM

The analysis in the last section comes from the general definition of the angular momentum operator. The next step is to consider specific physical cases. Dealing first with orbital angular momentum, the angular momentum due to the motion of a particle in a closed orbit, it is necessary to calculate the eigenfunctions $X(m, j)$. For orbital angular momentum, the symbols l and m_l are conventionally used in place of j and m.

Returning to the definition (B.1), and rewriting in spherical polar co-ordinates (r, θ, ϕ) it can be seen that L_z depends only on ϕ and so

$$X(l, m_l) = P(\theta) \exp i(m_l\, \phi). \tag{B.10}$$

It follows that exp $i(2\pi + \phi)$ must equal exp $i\phi$. Therefore m_l (and hence l) may only take integral values.

Solving (B.8) and (B.9), gives expressions for P which turn out to be a set of polynomials dependent on the quantum numbers l and m_l, and are known to mathematicians as spherical harmonics (Table B.1). These give the eigenfunctions of the orbital angular momentum of an electron. To repeat: in this case the eigenvalues are

$$\text{for } L_z : m_l\,(h/2\pi)$$

$$\text{for } |L|^2 : l\,(l+1)\,(h^2/4\pi^2)$$

where l can take the values 0, 1, 2, 3 etc., and, for a given l, the possible values of m_l are l, $(l - 1)$, $(l - 2)$ ....0....$(-l + 2)$, $(-l + 1)$, $-l$.

Table B.1 Spherical harmonics. $Y_{l,m}$ for values of l from 0 to 2.

$$Y_{0,0} = \frac{1}{\sqrt{4\pi}}$$

$$Y_{1,0} = \sqrt{\frac{3}{4\pi}}\cos\theta$$

$$Y_{1,1} = -\sqrt{\frac{3}{8\pi}}\,e^{i\phi}\sin\theta \qquad Y_{1,-1} = +\sqrt{\frac{3}{8\pi}}\,e^{-i\phi}\sin\theta$$

$$Y_{2,0} = \sqrt{\frac{5}{16\pi}}\,(3\cos^2\theta - 1)$$

$$Y_{2,1} = -\sqrt{\frac{15}{8\pi}}\,e^{i\phi}\sin\theta\cos\theta \qquad Y_{2,-1} = +\sqrt{\frac{15}{8\pi}}\,e^{-i\phi}\sin\theta\cos\theta$$

$$Y_{2,2} = -\sqrt{\frac{15}{32\pi}}\,e^{2i\phi}\sin^2\theta \qquad Y_{2,-2} = +\sqrt{\frac{15}{32\pi}}\,e^{-2i\phi}\sin^2\theta$$

B.3 SPIN ANGULAR MOMENTUM

As well as angular momentum due to orbital motion, the electron is found experimentally to have angular momentum due to 'spin', seen as motion about its own axis. It is found that the electron's angular momentum due to its spin can only take two values. So, from (B.6) and (B.7), the only possible values of j and m are

$$j = {}^1/_2, \quad m = + {}^1/_2 \text{ or } - {}^1/_2.$$

Commonly, for spin angular momentum the symbol j is replaced by s, and m by m_s (in Chapter 5, m_s is replaced by s for simplicity. This should not cause confusion)

This special case of (B.8) and (B.9) gives rise to only two eigenfunctions. One corresponds to $m_s = + {}^1/_2$, and is referred to as the 'spin-up' state, α. The other, which corresponds to $m_s = - {}^1/_2$ is referred to as the 'spin-down' state, β. Possible physical expressions for α and β cannot be discussed here.

The eigenequations then reduce to

$$\begin{aligned} |\mathbf{s}|^2\, \alpha &= {}^3/_4\, (h^2/4\pi^2)\ \alpha\, ; \quad |\mathbf{s}|^2\, \beta = {}^3/_4\, (h^2/4\pi^2)\ \beta \\ s_z\, \alpha &= {}^1/_2\, (h/2\pi)\ \alpha\ ; \quad s_z\, \beta = -{}^1/_2\, (h/2\pi)\ \beta. \end{aligned} \qquad \text{(B.11)}$$

B.4 ADDITION OF SPINS

The peculiarly simple mathematical relationship between the spin-up and spin-down eigenfunctions makes possible one very useful result when the total spin state of a pair of electrons (labelled 1 and 2) is required.

In this case, the total eigenfunction of the pair $\chi(1,2)$ is considered. The eigenequation for the square of the total spin amplitude $|\mathbf{S}|^2$ is given by

$$|\mathbf{S}|^2\, \chi(1,2) = S(S + 1)\, (h^2/4\pi^2)\, \chi(1,2). \qquad \text{(B.12)}$$

Now, since the electrons are independent particles, $\chi(1,2)$ must be made up of linear combinations of single-electron eigenfunctions, terms like $\alpha(1)\alpha(2)$, $\alpha(1)\beta(2)$, etc.

A little mathematical manipulation shows that the operator $|\mathbf{S}|^2$ may be written in terms of the spin operators for the individual electrons as follows

$$|\mathbf{S}|^2 = |\mathbf{s}_1|^2 + |\mathbf{s}_2|^2 + 2\, \mathbf{s}_{1z}\mathbf{s}_{2z} + \mathbf{s}_{1+}.\mathbf{s}_{2-} + \mathbf{s}_{1-}.\mathbf{s}_{2+}. \qquad \text{(B.13)}$$

The subscripts refer to the individual electrons; $\mathbf{s}_+$ and $\mathbf{s}_-$ are raising and lowering operators, as defined in section B.1.

Substitution in (B.12), using (B.13) shows that only four possible linear combinations satisfy the equation

$$
\begin{gathered}
\alpha(1)\alpha(2) \\
\beta(1)\beta(2) \\
[\alpha(1)\beta(2) + \beta(1)\alpha(2)] / \sqrt{2} \\
[\alpha(1)\beta(2) - \beta(1)\alpha(2)] / \sqrt{2}.
\end{gathered}
\tag{B.14}
$$

The first three states have quantum number $S = 1$ and are spatially symmetric – exchange of the electrons leaves $\chi(1,2)$ unchanged. The fourth has quantum number $S = 0$ and is spatially antisymmetric – exchange of electrons changes the sign of $\chi(1,2)$ (The factors of $\sqrt{2}$ are merely for normalization).

These properties will be seen to be important in Appendix F.

B.5 ADDITION OF SPIN AND ORBITAL ANGULAR MOMENTUM

Classical angular momenta add vectorially. Similarly, the total angular momentum of an electron must be the vector sum of its orbital and spin angular momenta. Generally, this means that the total quantum mechanical angular momentum operator , **J** can be defined as

$$\mathbf{J} = \mathbf{L} + \mathbf{S}$$

where **J** has the components

$$J_x = L_x + S_x\,,\quad J_y = L_y + S_y\,,\quad J_z = L_z + S_z.$$

Operators **L** and **S** operate on different variables and therefore commute. Given the separate commutation relations for **L** and **S**, it can easily be shown that **J** obeys the general eigenequations

$$J_z\, X(m_j, j) = m(h/2\pi)\, X(m_j, j) \tag{B.15}$$

$$|\mathbf{J}|^2\, X(m_j, j) = j(j+1)\,(h^2/4\pi^2)\, X(m_j, j), \tag{B.16}$$

Furthermore, it can be seen that

$$|\mathbf{J}|^2 = |\mathbf{L}|^2 + |\mathbf{S}|^2 + 2\mathbf{L}.\mathbf{S}.$$

Thus eigenvalues of the operator **L.S** are given by eigenvalues of the operator

$$|\mathbf{J}|^2 - |\mathbf{L}|^2 - |\mathbf{S}|^2$$

and have values

$$\tfrac{1}{2}\,[\,J(J+1) - L\,(L+1) - S\,(S+1)]\,(h^2/4\pi^2). \qquad \text{(B.17)}$$

This result is used in Section 5.3.

It should be noted finally that $|\mathbf{J}|^2$, J_z, $|\mathbf{L}|^2$ and $|\mathbf{S}|^2$ all commute with each other. Thus a set of eigenstates for the total angular momentum of a multi-electron system can be completely defined in terms of the set of quantum numbers (J, J_z, L and S)

However, it can also be seen that since $|\mathbf{L}|^2$, $|\mathbf{S}|^2$, L_z and S_z also all commute with each other, the same system can be defined in terms of a different set of eigenstates defined by the four quantum numbers (L, M, S and M_s). For obvious reasons, the first representation is known as the 'coupled representation' and the second the 'uncoupled representation'.

The coupled states can be written in terms of linear combinations of uncoupled states and this formulation is useful when detailed discussion of atomic coupling schemes is required.

B.6 CALCULATION OF THE LANDÉ g FACTOR

This factor, used in Section 11.2, gives a further example of the manipulation of angular momentum operators.

The problem is to work out the perturbation energy term for an atom with both orbital and spin angular momentum in a $\boldsymbol{B}$ field. From (11.4), the function required is

$$W' = \gamma\,(\boldsymbol{L} + g\boldsymbol{S})\,.\,\boldsymbol{B}. \qquad \text{(B.18)}$$

Using classical vectors, and making the approximation (good to 0.1%) that the electron g factor is 2, this may be written:

$$W' = \gamma\,(\boldsymbol{L} + 2\boldsymbol{S})\,.\,\boldsymbol{B}$$
$$= \gamma\,(\boldsymbol{J} + \boldsymbol{S})\,.\,\boldsymbol{B}.$$

Simple geometry enables this to be rewritten

$$W' = \gamma\,(1 + \boldsymbol{J}.\,\boldsymbol{S}\,/\,|\boldsymbol{J}|^2)\,\boldsymbol{J}\,.\,\boldsymbol{B}$$

$$= \gamma\,[\,1 + (\,|\boldsymbol{J}|^2 - |\boldsymbol{L}|^2 + |\boldsymbol{S}|^2)/2|\boldsymbol{J}|^2\,]\,\boldsymbol{J}\,.\,\boldsymbol{B}.$$

Converting into quantum operators, the quantity in the square brackets can be rewritten in terms of the eigenvalues of the various operators

$$W' = \gamma \{ 1 + [J(J+1) - L(L+1) + S(S+1)]/2J(J+1) \} \mathbf{J} \cdot B \qquad \text{(B.19)}$$

$$= \gamma g_L \mathbf{J} \cdot B$$

Hence the Landé g factor, g_L is given by

$$g_L = 1 + [J(J+1) - L(L+1) + S(S+1)]/2J(J+1).$$

Appendix C

Time-independent perturbation theory

Time-independent perturbation theory is a very useful technique for calculating corrections to known eigenfunctions due to small stationary potentials. There are two cases, depending on whether or not the eigenfunctions are energy degenerate.

C.1 NON-DEGENERATE STATES

If the energy eigenstates for a system with a simple potential function are known, the effect of a small extra potential can be calculated as follows.

Assume a system has TISE

$$\mathrm{H}\,\psi = E\,\psi. \tag{C.1}$$

The energy operator has the form

$$\mathrm{H} = \mathrm{H_0} + \lambda\,\mathrm{H'}$$

where $\mathrm{H_0}$ is an energy operator whose eigenfunctions are known and $\mathrm{H'}$ is a small extra potential energy operator. The constant λ is merely used to keep track of orders of approximation.

The unperturbed operator $\mathrm{H_0}$ has eigenequation

$$\mathrm{H_0}\,u_n = E_n\,u_n. \tag{C.2}$$

Now for the state produced when the perturbation is present, let

$$\psi = u_n + \lambda\,u' + \lambda^2 u'' + \ldots.$$

and

$$E = E_n + \lambda\,E' + \lambda^2\,E'' + \ldots.$$

where u' and E' define first-order corrections to the unperturbed values u_n and E_n; u'' and E'' second-order corrections etc.

Substitute in (C.1) and equate powers of λ.

Coefficients of λ^0 give the unperturbed equation

$$\mathrm{H_0}\, u_n = E_n\, u_n$$

coefficients of λ give

$$\mathrm{H_0}\, u' + \mathrm{H'}\, u_n = E_n\, u' + E'\, u_n. \tag{C.3}$$

This equation gives E', the first-order correction to the energy eigenvalue E_n.

To work out E', assume that the first-order correction term, u', can be made up of a linear combination of the unperturbed eigenfunctions (this may not seem obvious, but can be justified)

$$u' = \Sigma\, a_m u_m \tag{C.4}$$

where a_m are numerical coefficients and the summation is over all possible eigenstates defined by the quantum number m.

Substituting (C.4) in (C.3), and using (C.2), we obtain

$$\mathrm{H'}\, u_n = \Sigma\, (E_n - E_m)\, a_m u_m + E'\, u_n.$$

Premultiplying through by $u_n{}^*$, and integrating over all space gives

$$\int u_n{}^*\mathrm{H'}\, u_n\, \mathrm{dv}$$
$$= \Sigma(E_n - E_m)\, a_m \int u_n{}^* u_m \mathrm{dv} + E' \int u_n{}^* u_n \mathrm{dv}. \tag{C.5}$$

The orthonormality of eigenstates means that the integral $\int u_n{}^* u_m \mathrm{dv}$ is zero except where $m = n$, in which case the integral is unity. Therefore the first term on the right hand side of (C.5) is zero and

$$E' = \int u_n{}^*\mathrm{H'}\, u_n\, \mathrm{dv}. \tag{C.6}$$

This is the first-order correction, caused by the perturbation energy H', to the eigenvalue E_n. For example, if a correction is required for the small extra energy of the hydrogen energy state due to spin–orbit coupling, H' takes on the form of W_{so} in (5.10).

It is also possible to calculate the correction to the eigenfunction, but this is not needed for the present work. Knowing the first-order correction enables a return to (C.3) to work out a second-order correction from the equation for terms in λ^2, but this is rarely necessary.

Note that the operation which leads from (C.5) to (C.6) cannot be carried out if the unperturbed eigenstates are degenerate ($E_n = E_m$).

C.2 DEGENERATE STATES

In this case, the set of unperturbed eigenfunctions must be described by two quantum numbers (for example the principal quantum number and the angular momentum quantum number) and may be written $u_{n,k}$. For simplicity, these can be written $u_{0,k}$. Let the total number of degenerate eigenfunctions be r.

Although the perturbed eigenfunctions will not necessarily be degenerate they must be able to be described in terms of a linear combination of the unperturbed eigenfunctions. As before

$$\mathrm{H} = \mathrm{H}_0 + \lambda\, \mathrm{H}'$$

and

$$E = E_0 + \lambda\, E'_{0,i}.$$

However, in this case the expression for the ith eigenstate is

$$u_i = \phi_{0,i} + \lambda\, u'_{0,i}$$

where

$$\phi_{0,k} = \Sigma_k\; c_{i,k}\, u_{0,k}.$$

and $c_{i,k}$ are constant coefficients. The sum is over all r eigenfunctions.

Writing down the equation equivalent to (C.3), integrating and using the orthonormality of eigenfunctions, gives a set of simultaneous equations for the $c_{i,k}$ terms

$$\Sigma_m\, c_{i,m}\, [E'_{0,m}\, \delta_{k,m} \;\; - \;\; \int u_{0,k}\, \mathrm{H}'\, u_{0,m}\, \mathrm{dv}] = 0$$

where $\delta_{k,m}$ is the Dirac delta function

The only non-trivial solutions for this set of equations occur when the $(r{\times}r)$ determinant defined by the coefficients of the equations is zero

$$[\; \int u_{0,k}\, \mathrm{H}'\, u_{0,m}\, \mathrm{dv} \; - E'_{0,m}\delta_{k,m}] = 0 \qquad \text{(C.7)}$$

and this gives values for the first-order correction $E'_{0,k}$ to the energy of each eigenstate. For example, for a doubly-degenerate state, $r = 2$ and (C.7) becomes a (2×2) determinant, resolvable to a quadratic equation in E' and giving two solutions, $E'_{0,1}$ and $E'_{0,2}$.

Appendix D

Time-dependent perturbation theory – the calculation of the Einstein *B* coefficient

If the energy eigenstates have been calculated for a system, the effect of a small extra **time-varying** potential can be calculated by the technique of time-dependent perturbation theory.

D.1 THE PROBLEM

An atom with a set of energy eigenstates u_k interacts with a classical electromagnetic field. The motion of an electron in the atom is described by the time-dependent Schrödinger equation

$$\mathrm{H}\,\Psi = \mathrm{i}(\mathrm{h}/2\pi)\mathrm{d}\Psi/\mathrm{d}t. \tag{D.1}$$

The energy operator is made up of the time-independent part, H_o, which provides the set of stationary eigenstates u_k, and a time-dependent perturbation $\mathrm{H}'(\mathrm{t})$ due to the interaction between the electron and the field

$$\mathrm{H} = \mathrm{H}_o + \mathrm{H}'.$$

A general state function for the system, Ψ, can be made up from a linear combination of unperturbed eigenfunctions, now including time-dependent terms

$$\Psi = \sum_k a_k(t)\, u_k \exp -\mathrm{i}(2\pi E_k t\,/\mathrm{h}) \tag{D.2}$$

where a_k is the probability of the atom being detected in state k and the summation is over all eigenstates. Note that a_k is a function of time in this case.

If only two eigenstates u_i and u_f are considered, and at time $t = 0$ the system is in state i

$$a_\mathrm{i}(0) = 1 \text{ and } a_\mathrm{f}(0) = 0.$$

If, after some time T, $a_i(T) = 0$ and $a_f(T) = 1$ the system has undergone a transition from state i to state f. The problem is therefore to work out a general expression for $a_k(t)$.

D.2 TO WORK OUT AN EQUATION FOR $a_k(t)$

Returning to the general situation and substituting (D.2) in (D.1)

$$\Sigma a_k(\mathrm{H_o}+\lambda \mathrm{H}')u_k \exp -i(\omega_k t) = \Sigma[a_k E_k + i(h/2\pi)(da_k/dt)]u_k \exp -i(\omega_k t). \quad (D.3)$$

(For simplicity, $2\pi E_k/h$ has been replaced by ω_k.)

Consider a particular state u_m. Multiplying through (D.3) by u_m^* and integrating over all space means that all terms including $\int u_m^* u_k dv$ are zero unless $m = k$. Hence (D.3) reduces to

$$i(h/2\pi)\,(da_m/dt) \exp -i(\omega_m t)$$
$$= \Sigma a_k \left[\int u_m^* \mathrm{H}' u_k\, dv\right] \exp -i(\omega_k t). \quad (D.4)$$

Since H' < $\mathrm{H_o}$, we may approximate, replacing H' with λH' and putting

$$a_k = a_k^{(0)} + \lambda\, a_k^{(1)} + \lambda^2\, a_k^{(2)} + ...$$

where, as in Appendix C, λ merely defines the order of approximation.

Substituting in (D.4) and separating out the orders of approximation, it is found that to a zeroth order of approximation (coefficient λ^0),

$$da_m^{(0)}/dt = 0. \quad (D.5)$$

Hence $a_m^{(0)}$ = constant, which is what would be expected.

To first order (coefficient λ)

$$da_m^{(1)}/dt$$
$$= -\,(2\pi/ih)\, \Sigma a_k^{(0)} \left[\int u_m^*\, \mathrm{H}' u_k\, dv\right] \exp -i(\omega_k - \omega_m)t. \quad (D.6)$$

This equation gives a general first order approximation for the variation of a_m with time.

Returning to the special case of an atom with only two states i and f, (D.5) and (D.6) give

$$a_i^{(0)}(t) = 1$$

$$a_i^{(1)}(t) = -\,(2\pi/ih) \int_0^t \left[\int u_i^*\, \mathrm{H}'\, u_f\, dv\right] \exp -i\omega_T t\, dt \quad (D.7a)$$

$$a_f^{(0)}(t) = 0$$

$$a_f^{(1)}(t) = + (2\pi/ih) \int_0^t [\int u_f^* H' \, u_i \, dv] \exp -i\omega_T t \, dt \qquad (D.7b)$$

where $\omega_T = (\omega_f - \omega_i)$, and is, of course, equal to $2\pi\nu$, where ν is the frequency in Hz of the radiation emitted in the transition.

Second- and higher-order approximations may be calculated by a similar method, but they are rarely needed. It can be seen that the first-order approximation is good enough for times long compared to $1/\nu$ and of the order of magnitude of the lifetime of the excited state.

D.3 THE SPECIAL CASE OF AN ATOMIC ELECTRON INTERACTING WITH AN E.M. FIELD

To obtain a solution in a particular case, an expression for H' must be inserted in (D.7a) and (D.7b). For an atom interacting with an e.m. field, classical e.m. theory gives

$$H' = 2D \cos \omega t$$

where D is the electric dipole moment induced in the atom by the field and ω is the frequency of oscillation of the field (magnetic interaction is assumed negligible).

Substitution into (D.7) and integrating over time gives

$$a_f^{(1)}(t) = (2\pi D_{12}/ih)\left[\frac{\exp(i(\omega_T + \omega)) - 1}{i(\omega_T + \omega)} + \frac{\exp(i(\omega_T - \omega)) - 1}{i(\omega_T - \omega)}\right]. \qquad (D.8)$$

A new term D_{12} has been introduced, where

$$D_{12} = \int u_f^* D \, u_i dv.$$

This quantity is known as the quantum dipole moment matrix element. If a single atomic electron is considered, D can be thought of classically as the distortion caused by the external classical electric field. In the quantum picture D_{12} therefore includes the effect of the 'charge distributions' for the electron in its initial and final state. Note that if u_f and u_i have the same spherical symmetry, D_{12} is zero (see Section 6.4).

Usually $\omega_T \sim \omega$, so (D.8) may be simplified by assuming the first term in the square brackets is negligible compared to the second (this simplification is often known as the 'rotating wave approximation', since it is equivalent to replacing the sinusoidal function in H' by an exponential function)

$$a_f^{(1)}(t) = (2\pi D_{12}/ih)\left[\frac{\exp i(\omega_T - \omega) - 1}{i(\omega_T - \omega)}\right]. \qquad (D.9)$$

The probability of the atom being in state f after time t is therefore given by

$$P_f(t) = |a_f^{(1)}(t)|^2 = (16\pi^2/h^2)|D_{12}|^2 \frac{\sin^2[(\omega_T - \omega)t/2]}{(\omega_T - \omega)^2}$$

or

$$P_f(t) = (4\pi^2/h^2)\, t^2\, |D_{12}|^2\, (\sin\xi/\xi)^2 \qquad (D.10)$$

where $\xi = (\omega_T - \omega)t/2)$.

D.4 PHYSICAL INTERPRETATION

Equation (D.10) contains the function $(\sin\xi/\xi)$ which appears at first sight to become indeterminate at $\xi = 0$ (when $\omega_T = \omega$). More careful analysis shows that this function has the form of a delta function for values of t large compared with the period of the radiation. As usual in physics, delta functions must be handled with care in order to see their role in the physical system they are being used to model. In this case (as discussed in Section 6.5), the function acts like a 'narrow band filter'. If energy in the electromagnetic field is distributed continuously over a frequency range near to ω, and the atomic transition is also broadened about ω_T, the function 'picks out' the interaction for which $\omega = \omega_T$ so that energy is conserved in the transition.

Appendix E

Quantization of the electromagnetic field

In Chapter 6 the quantum nature of the electromagnetic field was briefly discussed. This concept is so important for modern optics and atomic physics that the outline of the analysis is presented in this appendix. More detailed discussion will be found in the works listed in Appendix I.

E.1 THE STEPS TO THE QUANTIZATION OF THE E.M. FIELD

The first step in the process is to decompose the field into normal modes. This process can be carried out in terms of the modes of a perfect laser cavity as discussed in Section 10.3, with the mirrors assumed infinitely large so that the monochromatic standing wave field in the z-direction is uniform in the x- and y-directions. However, the procedure can just as well be carried out with free-space travelling wave modes, so long as cyclical boundary conditions are defined.

The next step is to write down an expression for the energy stored in a single mode and show that this expression is identical in form to that for the energy of a simple harmonic oscillator, with the electric and magnetic field amplitudes E and B having the same formal relationship as the oscillator position and momentum variables x and p.

Classical electromagnetic theory gives an expression for the energy in the field mode as

$$H = \int \left(\tfrac{1}{2}\varepsilon_0 E^2 + \tfrac{1}{2}\mu_0 B^2 \right) \mathrm{dv} \qquad \text{(E.1)}$$

where the field is assumed to be in free space and the integration is over the mode volume.

Using the relationship between E and B given by Maxwell's free-space wave equation, (E.1) may be rewritten

$$H = \tfrac{1}{2}(P^2 + \omega^2 Q^2) \tag{E.2}$$

where

$$E = KQ,\ B = KP \ \text{ and } \ P = (\mu_0/kc^2)\mathrm{d}Q/\mathrm{d}t$$

ω is the angular frequency of the oscillation ($2\pi\nu$) and k is the wave number. The constant K is merely a function of the mode volume.

The relation between P and Q in (E.2) is identical to that between the classical variables p and x in the equivalent expression for the energy of a simple harmonic oscillator of angular frequency ω, so the dynamics of the field mode can be formally identified with those of the oscillator. Quantization then involves replacing P and Q with quantum operators P and Q.

In most quantum mechanics texts, the solution of Schrödinger's equation for the SHO is carried out in terms of 'creation' and 'annihilation' operators which are linear combinations of the p and x operators. These operators are non-Hermitian and so do not represent observables, but provide a simple and elegant method of solving the equation.

In the present case, if the operators P and Q are replaced by the two operators

$$a^- = (2h\nu)^{1/2}\,(\omega Q + i\,P)$$

$$a^+ = (2h\nu)^{1/2}\,(\omega Q - i\,P)$$

the quantized version of equation (E.2) can be rewritten

$$H = h\nu(a^+a + \tfrac{1}{2}). \tag{E.3}$$

It can be shown that although a^- and a^+ are non-Hermitian, the product a^+a^- is, and it can be seen that this operator defines the energy in the field mode. By analogy with the SHO, it can be seen readily that the energy in the mode is quantized in multiples of hν. The energy eigenequation for the operator defined in (E.3) gives eigenvalues

$$E_n = h\nu(n + \tfrac{1}{2})$$

where n is an integer.

E.2 ELECTRIC AND MAGNETIC FIELDS

In the quantum mechanical analysis of the SHO, the creation and annihilation operators merely provide means for solving Schrödinger's equation. For electromagnetic fields, they have a greater importance.

Returning to the original problem, the aim must be to write down quantum operators equivalent to the classical E and B field amplitudes. In fact when this is done, the electric field operator comes out as a linear combination of a^- and a^+

$$\mathrm{E} = K'[a^+ \exp(2i\pi\nu t) + a^-\exp(-2i\pi\nu t)]. \tag{E.4}$$

The constant K' includes any terms which define the spatial dependence of the mode and it should be noted that this is not changed by the quantization process. To oversimplify, the easiest picture of the process is of the replacement of a classical field with a quantum 'operator field' with the same spatial properties. A similar expression can be written down for B.

The problem is that the operator defined in (E.4) is not Hermitian, and hence not observable. However, this does not cause a difficulty when the detection of an electric field is considered. Rewriting (E.4) as

$$\mathrm{E} = \mathrm{E}^+ + \mathrm{E}^-$$

the process of detection of radiation can be thought of in terms of an interaction between a field and an atom where the atom rests in an initial eigenstate u_i and ends in a final eigenstate u_f. It is easily shown that the function R which defines the detection of the field is :

$$R = \int u_f \, \mathrm{E}^+ u_i \, \mathrm{dv}.$$

A little mathematical manipulation shows that this is equivalent to

$$R = \int u_i \, \mathrm{E}^+\mathrm{E}^- \, u_i \mathrm{dv}.$$

The product $\mathrm{E}^+\mathrm{E}^-$ is Hermitian, so this equation is equivalent to the semi-classical case (see Section 6.5)

$$R \propto |E|^2.$$

However, the connection of the electric and magnetic field operators themselves with classical measurable quantities is complex. It is touched on briefly in Section 6.9.

Appendix F

The Pauli exclusion principle

Consideration of the time-independent Schrödinger equation for two identical non-interacting particles in a potential field leads to a very important result.

If the two particles are labelled 1 and 2, the TISE for each particle may be written as

$$\begin{aligned} \mathrm{H}(1)u_a(1) &= E_a\, u_a(1) \\ \mathrm{H}(2)u_b(2) &= E_b\, u_b(2). \end{aligned} \tag{F.1}$$

Here H(1) and H(2) are the energy operators for the two particles and u_a and u_b are any two eigenstates of H.

But, considering the system as a whole, the total energy operator H(1,2) is given by

$$\mathrm{H}(1,2) = \mathrm{H}(1) + \mathrm{H}(2)$$

and the TISE is

$$\mathrm{H}(1,2)U(1,2) = (E_a + E_b)\, U(1,2) \tag{F.2}$$

where $U(1,2)$ is an eigenstate of the two-particle system.

Since the two particles are independent, $U(1,2)$ can be made up of linear combinations of terms like $u_a(1)u_b(2)$. It can be seen that (E.2) can be satisfied by solutions of the form

$$\begin{aligned} U(1,2) = {} & u_a(1)u_b(2) \\ \text{or } & u_b(1)u_a(2) \\ \text{or } & (1/\sqrt{2})(u_a(1)u_b(2) \pm u_b(1)u_a(2)). \end{aligned} \tag{F.3}$$

(The factor $(1/\sqrt{2})$ is merely required for normalization).

However, the fact that the particles are identical means that there is a further constraint on the form of $U(1,2)$. The system must be unchanged if the two particles are exchanged. What this implies mathematically may be worked out as follows.

Consider a new operator, the permutation operator P_{12} which merely exchanges the particles. As the particles are identical, the commutator

$$[P_{12}, H(1,2)] = 0.$$

Therefore $U(1,2)$ must be an eigenfunction of P_{12}, and so

$$P_{12}\, U(1,2) = \lambda U(1,2)$$

where λ is a constant. But if the permutation operator is applied twice

$$P_{12}(P_{12}\, U(1,2)) = \lambda(P_{12}\, U(1,2)) = \lambda^2\, U(1,2)$$

and, by the definition of the permutation operator

$$P_{12}(P_{12}\, U(1,2)) = U(1,2).$$

Therefore

$$\lambda^2 = 1 \text{ and } \lambda = \pm 1.$$

This implies that there can only be two possible forms for $U(1,2)$

$$\text{SYMMETRIC where } P_{12}\, U(1,2) = U(1,2)$$

$$\text{ANTISYMMETRIC where } P_{12}\, U(1,2) = -\, U(1,2).$$

Therefore, looking back at (F.3), the only possible forms of $U(1,2)$ which satisfy equation (F.2) are the symmetric solution

$$U(1,2) = (1/\sqrt{2})(u_a(1)u_b(2) + u_b(1)u_a(2)) \tag{F.4}$$

or the antisymmetric solution

$$U(1,2) = (1/\sqrt{2})\,(u_a(1)u_b(2) - u_b(1)u_a(2)) \tag{F.5}$$

It appears that symmetry or antisymmetry under particle exchange is a characteristic of the particles themselves and not something that can be arranged in making up the state. The generalization, first made by Pauli is that

(1) Systems consisting of combinations of identical particles which have half integral values of the spin quantum number (such as electrons with $s = 1/2$) are described by antisymmetric state functions, such as (F.5). The probability distribution of states of such particles is described by Fermi–Dirac statistics and the particles are called fermions.

(2) Systems consisting of combinations of identical particles which have integral values of the spin quantum number (such as α particles with $s = 0$) are described by symmetric state functions such as (F.4). The distribution of states of such particles is described by Bose–Einstein statistics and the particles are called bosons.

There is an important corollary of (F.5). For a system with two electrons in the same state (e.g. state a)

$$U(1,2) = 0$$

and so the state does not exist.

The generalization of this important result is known as the Pauli exclusion principle: 'In a system of two (or more) fermions, no two fermions can exist in the same eigenstate'.

For example, in an atom no two electrons can have the same values of the set of quantum numbers (n, l, m and s), where s defines the eigenvalue of the z-component of the electron spin angular momentum operator $\mathbf{s}_z$.

Appendix G

Eigenstates in helium

The two electrons in the helium atom provide a special case of the analysis given in Appendix F

The eigenfunction of each electron has a spatial part u and a spin part χ. Since these parts are eigenstates of operators which operate on different variables, the two parts are independent and the total eigenfunction is in the form of a product, $u\chi$.

Considering the two helium electrons (labelled 1 and 2), an eigenfunction for the whole system $\Psi(1,2)$ can be written down. By the Pauli exclusion principle, this must be antisymmetric

$$\Psi(2,1) = -\Psi(1,2). \tag{G.1}$$

Now, as in the case of the single electron, this state function can be divided into the product of two parts, the spatial part, $U(1,2)$, composed of a combination of the spatial eigenfunctions of the individual electrons $u(1)$ and $u(2)$ and the spin part $X(1,2)$ composed of a combination of their spin eigenfunctions

$$\Psi(1,2) = U(1,2)X(1,2). \tag{G.2}$$

Therefore although $\Psi(1,2)$ must be antisymmetric, this antisymmetry can be produced in two ways.

(1) by combining a symmetric function for $U(1,2)$ with an antisymmetric function for $X(1,2)$.
(2) by combining an antisymmetric function for $U(1,2)$ with a symmetric function for $X(1,2)$.

From Appendix F, the symmetric and antisymmetric forms for the combination of spatial eigenstates are given by

$$U(1,2) = (1/\sqrt{2})[u(1)u'(2) + u'(1)u(2)] \qquad \text{(symmetric)}$$

$$U(1,2) = (1/\sqrt{2})[u(1)u'(2) - u'(1)\,u(2)] \qquad \text{(antisymmetric)}.$$

Similarly, Appendix B.4 gives possible eigenfunctions for a combination of two spins:

$$\begin{aligned} X(1,2) &= \alpha(1)\alpha(2) \\ &\text{or } \beta(1)\beta(2) \\ &\text{or } (1/\sqrt{2})[\alpha(1)\beta(2) + \beta(1)\alpha(2)] \\ &\text{or } (1/\sqrt{2})[\alpha(1)\beta(2) - \beta(1)\alpha(2)] \end{aligned}$$

The first three of these are symmetric and the fourth antisymmetric.

Substituting into equation (G.2), having regard for the fact that $\Psi(1,2)$ must be antisymmetric, gives a 'singlet' state with total spin zero

$$\Psi(1,2) = [u(1)u'(2) + u'(1)u(2)]\,[\alpha(1)\beta(2) - \beta(1)\alpha(2)]/2 \qquad \text{(G.3)}$$

and three 'triplet' states, each with total spin unity:

:

$$\begin{aligned} \Psi(1,2) &= [u(1)u'(2) - u'(1)u(2)]\,\alpha(1)\alpha(2)/\sqrt{2} \\ \Psi(1,2) &= [u(1)u'(2) - u'(1)u(2)]\,\beta(1)\beta(2)/\sqrt{2} \qquad \text{(G.4)} \\ \Psi(1,2) &= [u(1)u'(2) - u'(1)u(2)]\,[\alpha(1)\beta(2) + \beta(1)\alpha(2)]/2. \end{aligned}$$

Thus there are four possible forms for the eigenstate $\Psi(1,2)$, three with a symmetric spin component and a total spin quantum number of 1, one with an antisymmetric spin component and a total spin quantum number of 0.

Appendix H

Exercises

Chapter 1

1 In a cathode ray tube, electrons are emitted from a heated cathode held at a potential of -1000 V and attracted towards an anode held at earth potential (see Fig. 1.1). Assuming that an electron has zero kinetic energy when it leaves the cathode. what is its kinetic energy (in joules) and speed (in m/s) when it reaches the anode?

2 In the CRT described in question 1, the electrons pass through a hole in the anode. The beam of electrons then travels between two plane parallel plates 0.05 m long separated by 0.01 m, and hits a fluorescent screen at a distance of 0.5 m from the ends of the plates. One plate is connected to a potential of +100V and the other to one of -100 V. How far and in which direction would the spot on the screen be deflected?

3 By considering how the electrons would be deflected by a uniform magnetic field, discuss how the CRT described in question 2 could be used to measure the charge/mass ratio of the electron.

4 In the CRT described in question 2, the cathode is replaced by an emitter of 7.7 MeV α particles and all electrical potentials are reversed. The fluorescent screen registers a pulse of light when an α particle hits it. Given that a particles are doubly ionized helium atoms, how much would the beam be deflected when electrical potentials are applied to the plates as in question 2?

5 In Millikan's experiment, droplets of oil fall vertically under gravity between two parallel horizontal plates, with the top plate at a positive potential relative to the bottom one. As the potential is changed, the rate of fall of the droplets is slowed or even reversed. How could observations of the motion of individual droplets be used to derive a value of the charge of an electron?

6 An electrolytic cell consists of a silver anode and a copper cathode immersed in a bath of silver nitrate. A current of 1 amp is passed through the cell for 30 minutes. When the cell is dismantled, it is found that a mass of 2.0 grams of silver has been deposited on the cathode. Assuming that the silver nitrate dissociates in solution into singly ionized silver ions and nitrate ions, use the known values of the charge and mass of the electron to work out the mass of a silver atom.

7 In an electrolytic cell containing dilute sulphuric acid, the passage of an electric current between the electrodes produces the emission of hydrogen gas at the cathode and oxygen at the anode. A current of 1 amp is passed for one hour and a mass of 0.75 g of hydrogen gas is collected. Work out the mass of the hydrogen atom.

8 Compare the results of questions 6 and 7 to work out the atomic weight of silver.

9 X-rays reflected from the surface of a sodium chloride crystal display a diffraction pattern rather similar to that seen when visible light is reflected from a diffraction grating. Assuming that sodium chloride has a simple cubic structure (with sodium and chlorine ions arranged alternately on a cubic lattice), estimate the wavelength of the X-rays, and compare it to the wavelength of visible light (the specific gravity of rock salt is 2.17).

10 Pure iron crystals have a 'body centred cubic' structure, consisting of cubic cells each of which has one iron atom at each corner and one in the middle of the cell. Given that the specific gravity of iron is 7.62 and its atomic weight is 55.8, estimate the distance between atoms in an iron crystal. (Hint: consider one cube and work out the number of **complete** atoms which make it up.)

Chapter 2

1 In Thomson's model, an atom consists of a sphere of positive charge with electrons embedded in it. Considering only the positive charge, the greatest electric field, and hence the greatest force on an α particle, will occur when the particle just grazes the surface of the atom. Given that the radius of an atom is approximately 10^{-10} m, apply (2.5) to work out the maximum angle to which a 7.7 MeV α particle could be deflected by a single gold atom (atomic number of gold = 79).

2 A beam of 7.7 MeV α particles is incident on a gold foil of thickness 2×10^{-7} m. Assuming Rutherford scattering, what proportion is scattered to an angle of 90° or more? (Specific gravity of gold = 19.3, atomic weight = 197).

3 In the scattering experiment described in question 2, a small but non-zero number of α particles is deflected to an angle of 180°. Use this fact to estimate a maximum size for the gold nucleus (assumed much larger than the α particle).

4 A beam of 8.3 MeV α particles is incident on a platinum foil. It is found that the Rutherford scattering formula ceases to be obeyed at scattering angles exceeding 90°. Assuming the α particle to have a radius of 2×10^{-15} m, estimate the radius of the platinum nucleus.

5 Use (2.9), (2.10) and (2.11) to work out an expression for the time that a classical electron in a hydrogen atom would take to spiral in from a distance of 10^{-10} m from the nucleus (the typical size of an atom) to a distance of 10^{-15} m (a typical nuclear size).

Chapter 3

1 Calculate the de Broglie wavelength of an electron in the CRT described in question 1.2.

2 In the CRT described in question 1.2, a thin film of graphite is mounted directly in front of the hole in the anode. Given that the specific gravity of graphite is 2.25 and the atomic weight of carbon is 12, discuss whether an electron diffraction pattern would be expected to be observed on the fluorescent screen.

3 Work out the de Broglie wavelength for a 7.7 MeV α particle.

4 Given the result of question 3, will quantum effects be noticeable in the Rutherford scattering experiment?

5 A quantum particle exists in a 1-dimensional well of length l, with infinitely high walls. Use the time-independent Schrödinger equation (3.10) to work out a general expression for the possible eigenenergies of the particle. What is the energy of the ground state? (Hint: the potential energy of the particle, V is zero for $0 < x < l$ and infinite for all other values of x.)

6 Sketch the eigenfunctions $U(x)$ for the three lowest eigenstates of the particle in problem 5, and hence the probability of finding the particle at any point.

7 The result of question 5 does not obey the correspondence principle. Justify this statement and explain the reason.

8 The ground state wave function for an electron in the hydrogen atom is

$$U_{100}(r) = (\pi r_o^3)^{-1/2} \exp -(r/r_o)$$
$$(r_o = h^2 \varepsilon_o/\pi m e^2).$$

Calculate the average distance of the electron from the nucleus and the most probable distance from the nucleus. Why are these different?

9 In the simple quantum mechanical model of the hydrogen atom, for any given value of the principal quantum number, n, there are a number of states with the same energy eigenvalue (called degenerate states). Use the relationship between the quantum numbers n, l and m to work out a relationship between n and the number of degenerate states. (Note that the inclusion of the spin quantum number s doubles the number of possible states.)

10 Use Table 3.1 to sketch the variation of U with θ and ϕ and hence the angular variation of the electron probability function for states in hydrogen with $n = 3, l = 2$, with $n = 3, l = 1$ and with $n = 3, l = 0$.

Chapter 4

1 A prism spectrometer for observing the near ultraviolet region of the spectrum around 200 nm, has a prism made of fused silica with a dispersion of 2.1 milliradians/nm. Use (4.1) to work out value for the prism aperture required for a resolving power of 10^5. Is this easily realizable in a real instrument?

2 In the visible spectrum of potassium, the line of longest wavelength turns out to be a doublet (λ_1 = 764.5 nm, λ_2 = 769.9 nm) Work out the dimensions required for a spectrometer using a prism made of flint glass (dispersion = 0.20 milliradians/nm) which can resolve these two lines.

3 A gas discharge tube contains helium at a pressure of 0.15 Pa and a temperature of 300 K. Estimate the mean free path of the atoms and the average time between collisions. How does this demonstrate that atoms in the gas discharge may be considered independent?

4 The discharge tube described in question 3 is excited by a current of 0.1 A between a cathode and an anode separated from it by 0.25 m and held at a potential of 1000 V(see Fig. 4.3). Estimate the mean free path of the electrons.

5 The first few lines of the visible series in the spectrum of atomic hydrogen lie at 656.46 nm, 486.27 nm, 434.17 nm and 410.29 nm. Use these values, together with the theoretical equation (4.5) to work out appropriate values of the quantum numbers n_1 and n_2, and the Rydberg constant.

6 In the discharge tube described in question 4, the discharge consists of a glowing 'positive column' and a dark space near the cathode. Use the results of question 5 to estimate the size of the dark space.

7 A chemist friend tells you that the spectrum of the one-electron ion of an element showed its 'ns' orbitals to be at energies of 0; 2057972 cm^{-1}; 2439156 cm^{-1} and 2572562 cm^{-1}. Convert into more comprehensible units, and identify the element.

8 Work out the energy required to remove the electron from the ion described in question 7.

9 A possible model of the potassium atom is a positive nucleus surrounded by a spherically symetrical sphere of negative charge, with a single electron in a 4s state outside this sphere. Given the ionization energy of potassium is 4.3 eV, what would be the effective atomic number Z^* as seen by the outer electron?

Chapter 5

1 Derive (5.2), by considering the electron as a classical massive sphere spinning about an axis, with uniform charge distributed throughout its volume.

2 Explain the inability of the picture of the electron as a classical spinning sphere of charge to explain the fine structure of atomic spectra.

3 Considering an electron in a 2p state in hydrogen, compare the magnitude of the spin–orbit energy term (W'_{so} in (5.10)) with the electron's potential energy due to electrostatic attraction to the nucleus.

4 Confirm that there are only two possible values for the z-component of the electron spin specified by quantum numbers +1/2 and -1/2.

5 Using (5.10) and (5.11), calculate the spin–orbit energy splitting of the 2p state in hydrogen.

6 Use (5.12) and (5.13) to derive (5.14), the frequency spread of an atomic line due to Doppler broadening .

7 Work out the Doppler-broadened width of the line at 568.3 nm in the sodium spectrum, if the gas of sodium atoms is at a temperature of 300 K.

8 A sodium discharge lamp is observed, using the spectrometer described in Section 4.2. How hot would the gas have to be before it was impossible to resolve the two sodium lines at 589.0 nm and 589.6 nm?

9 The natural line width of a transition in sodium is calculated to be 5×10^7 Hz. At what gas temperature would the Doppler-broadened line width be of the same magnitude as the natural line width?

Chapter 6

1 Given the atom/field system described in Section 6.2, use (6.8) and (6.9) to work out the simple relationship between the coefficients of spontaneous and stimulated emission

$$A_{21}/B_{21} = 8\pi h\, \nu_{12}^3/c^3 \quad \text{Js/m}^3.$$

2 An atom has three energy states 1, 2 and 3 ($E_1<E_2<E_3$). Transitions between all states are allowed. A gas of such atoms is subjected to a radiation field which has a high energy density only at frequencies close to $(E_3 - E_1)/h$. Draw an energy level diagram showing all possible transitions between states.

3 For the gas in question 2, show that in equilibrium the relative populations of states 2 and 3 are given by $N_2 = N_3\, A_{32}/A_{21}$.

4 For the gas in question 2, $A_{21} < A_{32}$. Show that it is possible for there to be an equilibrium situation where $N_2 > N_1$.

5 An atomic absorption experiment is carried out in which light from a tungsten filament is focused through a helium discharge onto the slit of a spectrometer. The temperature of the filament can be calculated from its resistance. At a filament temperature of 1500 K, a continuous spectrum with a dark line at 447.1 nm is seen. The filament temperature is decreased and at a temperature of 1200 K the dark line vanishes. At lower temperatures a bright line is seen at the same wavelength, superimposed on the continuous spectrum. What is the temperature of the gas? (Hint: write down equilibrium equations for atoms in the upper and lower states.)

6 Using (6.5) show that the value of the quantum dipole moment matrix element for a transition between states 1 and 2 is zero when u_1 and u_2 are both s states of a hydrogen-like atom.

7 Use symmetry arguments to extend the result of question 6 to transitions between any two states of the same angular symmetry.

8 As stated in Section 6.6, a reasonable approximation for the quantum dipole moment matrix element for an allowed optical transition is $-er_o$ Cm, where r_o is the Bohr radius. Use (6.10) and (6.19) to verify this for the 2p – 1s transition in hydrogen.

9 Use the arguments deployed in Section 6.5 to produce Fermi's 'golden rule' for the case of an atom ionized by an electromagnetic field. Hence derive the Einstein photoelectric equation.

10 Follow through the argument outlined in Section 6.6 to demonstrate that for dipole radiation, only transitions in which the m quantum number increases or

decreases by unity are allowed.

11 Sodium atoms in a particular state are excited by electron collision to a state of higher energy and are observed to rest in that state for an average time of 4×10^{-8} s before decaying back to their original state via spontaneous emission. Estimate the natural linewidth of this radiation, and compare it with the Doppler broadening discussed in question 5.7.

Chapter 7

1 By neglecting electron–electron interaction, show that (7.1) can be separated into two 'hydrogen-like' equations.

2 A system consists of three electrons each of which may be in eigenstate u_a, u_b, or u_c. Show that an acceptable eigenstate for the whole system is given by a 3×3 determinant. Show how this demonstrates the Pauli exclusion principle.

3 Write down and explain the electron configuration for the ground state of beryllium.

4 Write down and explain the ground state electron configuration for magnesium.

5 Write down the electron configurations of the ground states of the 'noble gases', neon, argon and krypton, and explain their chemical similarities.

6 Explain why the electron of highest energy in the ground state electron configuration of potassium is in a 4s state, not a 3d state.

7 Write down and explain the ground state electron configuration of caesium.

8 Continue the graph shown in figure 7.2 for values of atomic number greater than 50. Explain any maxima and minima in terms of the configurations of the electrons involved.

Chapter 8

1 Explain the way in which the eigenstates of helium divide into 'singlet' and 'triplet' states. By referring to Chapter 6, explain why there are no optically allowed singlet–triplet transitions.

2 Explain why the configurations (2s2s) and (2s2p) do not occur in the helium energy level diagram. (Hint: work out an approximate value for the energy.)

3 Consider the 2^1S, 2^1P, 2^3S and 2^3P states in helium. Starting from a sketch of the energy level diagram which neglects electron–electron energies, sketch in the effect of Coulomb and interaction energies in each case.

4 Use (8.6) and the hydrogen-like wave functions to write down an equation to give values for the interaction and exchange energies for the 1s1s $^1S_0$ state in helium.

5 Estimate the energy difference due to electron–electron interaction between the $1s2s^1S$ and the $1s2s^3S$ states in helium.

6 Write down and explain the term symbol for the ground states of lithium, beryllium and boron.

7 By considering the fine structure of the excited states of sodium and the electric dipole selection rules, show that lines in the sharp (mS–2P) and principal (mP–1S) series are doublets, while lines in the diffuse (mD–2P) series are triplets (see Fig. 9.1).

8 Consider the following state terms (not necessarily occurring in the same atom)

$$^1D,\ ^2P,\ ^4F,\ ^3G.$$

In each case, work out what possible values of J are implied, and list the possible levels that may arise.

9 For each of the sets of levels written down in question 7, use Hund's rules to identify the state of lowest energy.

Chapter 9

1 Write down and explain the ground state electron configuration of carbon. The configuration ends up with two electrons 'floating' outside closed shells. Use the fact that the wave function describing the two electrons must be antisymmetric, and Hund's rule to show that the ground state must be a ^3P state.

2 Write down the ground state electron configuration for boron ($Z = 5$). Given that the ionization energy of boron is 8.3eV calculate the effective 'screened' charge of the nucleus seen by the outermost electron.

3 The screening effect discussed in question 2 is produced by the inner closed shells of electrons. Considering these electrons as classical clouds of charge surrounding the nucleus, calculate the ionization energy if only screening by 1s electrons is considered. Calculate the value due to screening by both 1s and 2s electrons. Compare the two values with the experimental value.

4 Work out the values of the effective atomic number Z^* for ground state outer electrons in lithium, sodium and potassium.

5 Using the values calculated in the previous question, use (9.1) to plot the r dependence of the ground state Slater orbitals in the elements mentioned and compare them with the equivalent hydrogen-like eigenfunctions.

6 Write down the Hartree–Fock equation (9.3) for the special case of helium and show that it reduces to the Schrödinger equation for helium discussed in Chapter 8.

7 Show that two 'p' electrons coupled together via jj coupling may produce three states defined by $(j_1,j_2) = (^1/_2,^1/_2)$, $(^1/_2,^3/_2)$ or $(^3/_2,^3/_2)$ in increasing order of energy.

8 Elements in Group IV of the periodic table possess a ground state consisting of two 'p' electrons outside a closed shell. The lightest elements obey LS coupling, while the heaviest obey almost perfect jj coupling. Draw a sketch to show the relative energies of the different terms as a function of Z.

Chapter 10

1 A laser consists of a tube of gas 1 m long of cross section area 1 cm^2 with plane parallel mirrors of reflectivity 0.99 at each end. The laser transition is at 500 nm and the spontaneous lifetime for this transition ($1/A_{mn}$) is 10^{-8} s. By considering the gains and losses of a travelling wave on a 'round trip' between the mirrors, calculate in atoms per cubic centimetre, the population inversion required for amplification at the laser frequency.

2 In neon, the first excited state is 135 000 cm^{-1} above the ground state. Calculate the number of atoms per cubic centimetre in the ground state and in the excited state for a gas at a temperature of 300 K and pressure of 10 Pa. Compare this number with the result of question 10.1.

3 In the system described in question 10.1, what will be the frequency spacing of the resonant modes? Sketch the spectrum of resonant modes and superimpose the natural lineshape of the laser transition.

4 Assuming that the laser transition is in a neon discharge at a temperature of 300 K, draw the resonant mode spectrum and superimpose the Doppler broadened spectral lines. Estimate how many different modes will be excited by the laser operation.

5 For the transition described in question 10.4, the cavity resonant mode is at a frequency ν_l which is 2×10^7 Hz lower than ν_o, the resonant frequency of the transition. Describe the way in which this mode interacts with two separate sets of neon atoms

6 Write down the ground state configuration of singly-ionized argon. Three groups of levels are important for laser operation. In order of increasing energy, they are the $4s^2P$ levels, the $4p^4D$, $4p^2D$, $4p^2P$ and $4p^2S$ levels (see Fig. 10.8). Identify the values of J for these levels and hence the allowed transitions.

7 In the argon ion energy level diagram, the lifetime for spontaneous emission for the transition from the lowest excited state to the ground state is of the order of 10^{-10} s and the lifetimes for the other transitions are all of the order of 10^{-8} s. Use (10.21) to (10.24) to estimate the relative rates of excitation of two of the optically connected levels, which would be required for laser action to take place.

8 Given the laser system described in question 10.1, and assuming that when switched on, the excitation process produces an instantaneous population inversion twice that of the laser threshold, use (10.21) to (10.24) to estimate the energy density of the steady-state laser field in the cavity.

9 From the result of question 8, calculate the intensity of the laser beam emitted through either mirror.

10 Calculate the passive cavity Q of the laser system described in question 1 and hence the resonant linewidth. From the result of question 8, use (10.10) to estimate the frequency spread of a single laser mode.

11 A laser emits radiation in three adjacent modes, specified by the equation:

$$E_n = A \sin(\nu_n t + \phi_n(t))\exp(-ik_n z)$$

where $\nu_n = nc/2d$.

Work out the resultant field when $\phi_n(t)$ is

(a) a randomly varying function of t,

(b) a constant, independent of n.

Chapter 11

1 The Stern–Gerlach experiment was originally carried out using silver atoms and produced a splitting into two beams. Assuming it were technically possible, what result would be obtained if the experiment were repeated using a beam of (a) hydrogen atoms, (b) helium atoms, (c) carbon atoms, (d) oxygen atoms?

2 The lines in the Stern–Gerlach experiment are not infinitesimally narrow. What is the source of their broadening?

3 Sketch the Zeeman splitting for the $2^2P_{3/2}$–$1^2S_{1/2}$ transition in potassium. The transition produces radiation at a wavelength of 764 nm. If the potassium discharge tube is in a B field of 1 T, calculate the magnitude of the Zeeman splitting which would be observed.

4 A transition occurs between two singlet levels for which $J = 3$ and $J = 2$ respectively. Sketch the Zeeman splitting of these two levels and use selection rules to determine how many Zeeman components would be observed.

5 Use the result of degenerate time-independent perturbation theory equation (C.7) to calculate the Stark effect splitting between the 2s and 2p states in hydrogen due to an electric field E V/m.

6 Derive (11.7) and demonstrate that the solution of these equations is an oscillatory one.

7 Given the definition of the Rabi frequency, (11.9), work out its value as a function of electromagnetic field intensity for the Lyman α transition in hydrogen.

Chapter 12

1 An electron in a Penning trap oscillates in an electric field defined by (12.1). If the electron emits radiation at a frequency of 60 MHz, calculate the value of the constant A. Given the dimensions of the trap indicated in Fig. 12.1, estimate the distance over which (12.1) is valid.

2 Compare the operation of 'optical molasses' with the interaction of a standing wave field with atoms in a gas laser.

3 Does the 'watched pot' experiment constitute a paradox?

Answers to numerical problems

Chapter 1

1 1.6×10^{-16} J, 1.88×10^{7} m/s
2 2.4×10^{-2} m
4 6.4×10^{-5} m
6 1.77×10^{-25} kg
7 1.67×10^{-27} kg
8 106
9 of the order of 10^{-10} m
10 3.4×10^{-11} m.

Chapter 2

1 0.017°
2 0.002%
3 2.9×10^{-14} m
4 11.5×10^{-15} m
5 1.6×10^{-8} s

Chapter 3

1 3.8×10^{-11} m
3 5.2×10^{-15} m
8 5.3×10^{-11} m, 7.95×10^{-11} m
9 Degeneracy = n^2 (or $2n^2$ if spin is included)

Chapter 4

1 4.8 cm
2 Prism aperture >7 mm. Focal length defined by linear resolution of detector.
3 Approx. 0.25 m (depending on assumptions)
4 Approx. 0.25 m (depending on assumptions)
5 $n_1=2$, $n_2 = 3,4$ or 5; 2.2×10^{-18} J
6 1.65 cm
7 $Z=5$(boron).
8 340 eV
9 2.25

Chapter 5

3 Approx. 10^{-6}
5 7.25×10^{-24} J (4.53×10^{-5} eV)
7 1.44×10^{9} Hz
8 1.8×10^{8} K
9 1.8 K

Chapter 6

5 1200 K
11 3.98×10^{6} Hz

Chapter 7

3 $(1s)^2(2s)^2$
4 $(1s)^2(2s)^2(2p)^6(3s)^2$
5 $(1s)^2(2s)^2(2p)^6$, $(1s)^2(2s)^2(2p)^6(3s)^2(3p)^6$, etc.
7 (Xe)(6s)

Chapter 8

5 Approx. 0.9 eV
8 (2,1,0);
(3/2,1/2);
(9/2,7/2,5/2,3/2,1/2);
(5,4,3,2,1)
9 (2,1,0); (1,0); $^3P_0$

Chapter 9

1 $(1s)^2(2s)^2(2p)^2$
2 $(1s)^2(2s)^2(2p)$, 1.56
3 30.6 eV, 3.4 eV, 8.3 eV

Chapter 10

1 5×10^3 atoms/cm^3
2 exp(-650)
3 150 MHz
7 $R_2 > 100R_1$
8 1.3×10^{-13} J/m^3
9 3×10^{-5} W.
10 3600 Hz

Chapter 11

1 Splitting into 2, 0, 0
and 5 respectively.
3 1.41×10^{10} Hz

Chapter 12

1 6.4×10^4 V/m^2, approx. $\pm$0.5 mm
from centre.

Appendix I

Further reading

The following is not an exhaustive set of references, but is intended to give students somewhere to start if they wish to go further into a particular topic. Not all the books cited are in print, but they should all be available from a good library.

Chapter 1

Anderson, D.L. (1964) *The Discovery of the Electron*, Van Nostrand, Princeton.
Keller, A. (1983) *The Infancy of Atomic Physics*,, Clarendon Press, Oxford.

Chapter 2

Keller, A. (1983) *The Infancy of Atomic Physics*, Clarendon Press, Oxford.
Shamos, M.H. (1959) *Great Experiments in Physics*, Holt, Rinehart and Winston, New York.

Chapter 3

Davies, P.C.W. and Betts, D.S. (1994) *Quantum Mechanics*, 2nd edn, Chapman & Hall, London.
Rae, A. (1992) *Quantum Mechanics*, 3rd edn, IOP Publishing, Bristol.
Gasiorowicz, S. (1996) *Quantum Physics*, 2nd edn, Wiley, New York.

at a higher level

Ballentine, L.E. (1990) *Quantum Mechanics*, Prentice Hall International, New Jersey.

Chapter 4

Kuhn, H.G. (1964) *Atomic Spectra,* Longmans, London.
Thorne, A.P. (1988) *Spectrophysics,* Chapman & Hall, London.
Herzberg, G. (1944) *Atomic Spectra and Atomic Structure,* Dover, New York.

Chapter 5

Atkins, P.W. (1983) *Molecular Quantum Mechanics,* 2nd edn, Clarendon Press, Oxford.

Chapter 6

Davies and Betts *op. cit.*
Gasiorowicz *op. cit.*
Knight, P.L. and Allen, L. (1983) *Concepts of Quantum Optics,* Pergamon, Oxford.
Loudon, R. (1983) *The Quantum Theory of Light,* 2nd edn, Clarendon Press, Oxford.

Chapter 7

Gasiorowicz *op. cit.*
Atkins *op. cit.*

Chapter 8

Atkins *op. cit.*

Chapter 9

Atkins *op. cit.*

Chapter 10

Meystre, P. and Sargent, M. (1990) *Elements of Quantum Optics,* Springer -Verlag, Berlin.
Milonni, P.W. and Eberley, J.H. (1988) *Lasers,* Wiley, New York.
Sargent, M,, Sculley, M.O. and Lamb, Willis E.(1974) *Laser Physics,* Addison Wesley, London.

Chapter 11

Atkins *op. cit.*
Herzberg *op. cit.*
Smith, K.F. (1955) *Molecular Beams*, Methuen, London.
Pendlebury, J.M. and Smith, K.F. (1987) Molecular beams, *Contemp. Phys.* **28**(1), 3–32

Chapter 12

von Bayeren, H.C.(1994) *Taming the Atom*, Penguin, London.
Dehmelt, H. (1990) Less is more: Experiments with atomic particles at rest, *Amer. J. Phys.* **58**(1),17–27.
Itano, W.M., Heinzen, D.J., Bollinger, J.J. and Wineland, D.J. (1990) Quantum zeno effect, *Phys. Rev. A.* **41**, 5, 2295–2300.
Stenholm, S. (1988) Light forces put a handle on the atom: To cool and trap atoms by laser light, *Contemp. Phys.* **29**(2) 105–123

at a higher level

Berman, P.R. (1994) *Cavity Quantum Electrodynamics*, Academic Press, Boston.

Appendix J

Useful constants

General

Speed of light in a vacuum	$c = 3.000\times10^{8}$ m/s
Electric permittivity of free space	$\varepsilon_0 = 8.854\times10^{-12}$ C²/Nm²
	$1/4\pi\varepsilon_0 = 8.988\times10^{9}$ Nm²/C²
Magnetic permeability of free space	$\mu_0 = 1.257\times10^{-6}$ Tm/A
Planck's constant	$h = 6.626\times10^{-34}$ Js
Boltzmann's constant	$k = 1.381\times10^{-23}$ J/K

Electronic

Electron charge	$e = 1.602\times10^{-19}$ C
Electron rest mass	$m = 9.109\times10^{-31}$ kg
Electron-Volt	1eV= 1.602×10^{-19} J
Bohr magneton	$\mu_b = eh/4\pi m = 9.274\times10^{-24}$ J/T

Atomic

Avogadro's number	$N = 6.022\times10^{23}$ mol^{-1}
Rest mass of proton	$m_p = 1.673\times10^{-27}$ kg
Atomic weight of hydrogen (defined with C_{12}= 12)	$M = 1.008$
Rydberg constant	$R = 1.097\times10^{7}$ m^{-1}
Bohr radius	$r_0 = 0.529\times10^{-10}$ m

Index

Printed in the United States
by Baker & Taylor Publisher Services